# Asymmetric and Selective Biocatalysis

# Asymmetric and Selective Biocatalysis

Special Issue Editors

**Jose M. Palomo**
**Cesar Mateo**

MDPI • Basel • Beijing • Wuhan • Barcelona • Belgrade

**MDPI**

*Special Issue Editors*

Jose M. Palomo
Group of Chemical Biology and Biocatalysis
Departament of Biocatalysis
Institute of Catalysis (ICP-CSIC)
Spain

Cesar Mateo
Group of Chemical Processes Catalyzed by enzymes
Departament of Biocatalysis
Institute of Catalysis (ICP-CSIC)
Spain

*Editorial Office*
MDPI
St. Alban-Anlage 66
4052 Basel, Switzerland

This is a reprint of articles from the Special Issue published online in the open access journal *Catalysts* (ISSN 2073-4344) from 2016 to 2017 (available at: https://www.mdpi.com/journal/catalysts/special_issues/selective_biocatalysis)

For citation purposes, cite each article independently as indicated on the article page online and as indicated below:

LastName, A.A.; LastName, B.B.; LastName, C.C. Article Title. *Journal Name* **Year**, *Article Number*, Page Range.

**ISBN 978-3-03897-846-6 (Pbk)**
**ISBN 978-3-03897-847-3 (PDF)**

# Contents

# About the Special Issue Editors

**Jose M. Palomo** was born in Coin, Malaga, Spain (1976). He received his bachelor's degree in Chemistry in 1999 from the Universidad Autonoma de Madrid (UAM) as a Spanish government fellow. After graduation, he joined Guisan's group at Institute of Catalysis (ICP, CSIC) in Madrid to complete his doctoral studies. He received his Ph.D. degree (summa cum laude) in 2003. His Ph.D. thesis developed novel highly efficient immobilized biocatalysts for asymmetric transformations and strategies for understanding the biochemical features of lipases. As an EMBO postdoctoral fellow, Jose pursued training in solid-phase peptide synthesis and semisynthetic protein preparation with Herbert Waldmann at the Chemical Biology Department of the Max Planck Institute in Dortmund from 2004 to 2006. In 2006, he began his appointment as Associate Research Scientist at Biocatalysis Department in ICP-CSIC. From 2009, he has been an Associate Professor (Tenured Scientist) in ICP-CSIC. Jose has published more than 140 articles in high-impact journals, with more than 7000 times cited and an H index of 42.

**Cesar Mateo** was born in 1969 and he did his PhD in the Catalysis Institute (ICP-CSIC), receiving the title in 2001 in the Autonoma University (Madrid, Spain). After a period of study in Marseille with Prof. Furstoss and in Delft with Prof. Sheldon, he started working in ICP. He is an Associate Professor (Tenured Scientist) in the ICP. He has focused his career in developing different immobilization technologies to obtain heterogeneous biocatalysts with improved properties. The catalysts have been used for different biotransformations, developing sensors (based on enzymes, antibodies, or DNA probes), or in other fields such as in biomedical applications and others. As principal contributions, he is the author of 164 articles in SCI journals, more than 20 book chapters, and 14 patents, with an H index of 53. The papers have been cited more than 9800 times in different journals.

# Preface to "Asymmetric and Selective Biocatalysis"

The synthesis of compounds or chiral building blocks with the desired configuration is one of the greatest challenges of chemistry, and is of the great interest in fields such as analytical chemistry and especially in fine and pharmaceutical chemistry. For this, different biocatalysts (i.e., cells, enzymes, catalytic antibodies, or ribozymes) have been used to catalyze different processes used, even on an industrial scale. Biocatalysts have a high activity under very mild conditions, such as ambient temperature, neutral pH, and atmospheric pressure. They are also able to catalyze highly selective and specific modifications in different substrates with high complexity, allowing the synthesis of enantiomerically pure compounds either by resolution processes or by asymmetric synthesis from prochiral substrates or regioselective modifications in complex molecules. This avoids side reactions as well as costly purification processes.

In addition to the pure biocatalysts that are traditionally used, in recent years, different hybrid catalysts have been developed that combine the good catalytic properties of traditional biocatalysts with the properties of organometallic catalysts. In this way, different mixed catalysts have been developed as artificial metalloenzymes combining enzymatic and metallic catalytic activities, expanding the applicability to different systems, such as cascade processes.

**Jose M. Palomo, Cesar Mateo**
*Special Issue Editors*

# catalysts

MDPI

*Editorial*

# Asymmetric and Selective Biocatalysis

## Cesar Mateo * and Jose M. Palomo *

Department of Biocatalysis, Institute of Catalysis (ICP-CSIC), Marie Curie 2, Cantoblanco, Campus UAM, 28049 Madrid, Spain
* Correspondence: ce.mateo@icp.csic.es (C.M.); josempalomo@icp.csic.es (J.M.P.);
  Tel.: +34-915-854-768 (C.M. & J.M.P.)

Received: 14 November 2018; Accepted: 15 November 2018; Published: 28 November 2018

The synthesis of compounds or chiral building-blocks with the desired configuration is one of the greatest challenges of chemistry and is of great interest in different fields such as analytical chemistry and especially in fine and pharmaceutical chemistry. Different biocatalysts (cells, enzymes, catalytic antibodies, or ribozymes) have been used to catalyze different processes, even on an industrial scale. Biocatalysts have high activity under very mild conditions such as ambient temperature, neutral pH, and atmospheric pressure. They are also able to catalyze highly selective and specific modifications in different substrates with high complexity, allowing the synthesis of enantiomerically pure compounds either by resolution processes or by asymmetric synthesis from prochiral substrates or regioselective modifications in complex molecules. This avoids side reactions as well as costly purification processes.

In recent years, in addition to the pure biocatalysts traditionally used, different hybrid catalysts have been developed, which combine the good catalytic properties of traditional biocatalysts with the properties of organometallic catalysts. In this way, different mixed catalysts have been developed as artificial metalloenzymes combining enzymatic and metallic catalytic activities, expanding the applicability to different systems such as cascade processes.

This issue contains one communication, six articles, and two reviews.

The communication from Paola Vitale et al. [1] represents a work where whole-cells were used as biocatalysts for the reduction of optically active chloroalkyl arylketones, followed by a chemical cyclization to give the desired heterocycles. Among the various whole cells screened (baker's yeast, *Kluyveromyces marxianus* CBS 6556, *Saccharomyces cerevisiae* CBS 7336, *Lactobacillus reuteri* DSM 20016), baker's yeast provided the best yields and the highest enantiomeric ratios (95:5) in the bioreduction of the above ketones. In this respect, valuable, chiral non-racemic functionalized oxygen-containing heterocycles (e.g., (S)-styrene oxide, (S)-2-phenyloxetane, (S)-2-phenyltetrahydrofuran), amenable to be further elaborated on, can be smoothly and successfully generated from their prochiral precursors.

Research regarding pure biocatalysts utilizing mechanistic studies, their application in different reactions and new immobilization methods for improving stability featured in five different articles.

The article by Su-Yan Wang et al. [2] describes the cloning, expression, purification, and characterization of an *N*-acetylglucosamine 2-epimerase from *Pedobacter heparinus* (PhGn2E). For this research, several *N*-acylated glucosamine derivatives were chemically synthesized and used to test the substrate specificity of the enzyme. The mechanism of the enzyme was studied by hydrogen/deuterium NMR. The study of the anomeric hydroxyl group and C-2 position of the substrate in the reaction mixture confirmed the epimerization reaction via ring-opening/enolate formation. Site-directed mutagenesis was also used to confirm the proposed mechanism of this interesting enzyme.

The article by Forest H. Andrews et al. [3] studies two enzymes benzoylformate decarboxylase (BFDC) and pyruvate decarboxylase (PDC) that catalyze the non-oxidative decarboxylation of 2-keto acids with different specificity. BFDC from *P. putida* exhibited very limited activity with pyruvate, whereas the PDCs from *S. cerevisiae* or from *Z. mobilis* showed virtually no activity with benzoylformate (phenylglyoxylate). After research using saturation, mutagenesis BFDC T377L/A460Y variant was obtained, with a 10,000-fold increase in pyruvate/benzoylformate. The change was attributed to an improvement in the *Km* value for pyruvate and a decrease in the $k_{cat}$ value for benzoylformate. The characterization of the new catalyst was performed providing context for the observed changes in the specificity.

The article by Xin Wang et al. [4] compares two types of biocatalysts to produce D-lysine L-lysine in a cascade process catalyzed by two enzymes: racemase from microorganisms that racemize L-lysine to give D,L-lysine and decarboxylase that can be in cells, permeabilized cells, and the isolated enzyme. The comparison between the different forms demonstrated that the isolated enzyme showed greater decarboxylase activity. Under optimal conditions, 750.7 mmol/L D-lysine was finally obtained from 1710 mmol/L L-lysine after 1 h of racemization reaction and 0.5 h of decarboxylation reaction. D-lysine yield could reach 48.8% with enantiomeric excess (ee) of 99%.

In the article of Rivero and Palomo [5], lipase from *Candida rugosa* (CRL) was highly stabilized at alkaline pH in the presence of PEG, which permits its immobilization for the first time by multipoint covalent attachment on different aldehyde-activated matrices. Different covalent immobilized preparation of the enzyme was successfully obtained. The thermal and solvent stability was highly increased by this treatment and the novel catalysts showed high regioselectivity in the deprotection of per-*O*-acetylated nucleosides.

The article by Robson Carlos Alnoch et al. [6] describes the protocol and use of a new generation of tailor-made bifunctional supports activated with alkyl groups that allow the immobilization of proteins through the most hydrophobic region of the protein surface and aldehyde groups that allow the covalent immobilization of the previously adsorbed proteins. These supports were especially used in the case of lipase immobilization. The immobilization of a new metagenomic lipase (LipC12) yielded a biocatalyst 3.5-fold more active and 5000-fold more stable than the soluble enzyme. The PEGylated immobilized lipase showed high regioselectivity, producing high yields of the C-3 monodeacetylated product at pH 5.0 and 4 °C.

The hybrid catalysts composed by an enzyme and metallic complex is also covered in this Special Issue.

The article by Christian Herrero et al. [7] describes the development of the Mn(TpCPP)-Xln10 A artificial metalloenzyme, obtained by non-covalent insertion of Mn(III)-meso-tetrakis(p-carboxyphenyl)porphyrin [Mn(TpCPP), 1-Mn] into xylanase 10 A from *Streptomyces lividans* (Xln10 A). The complex was found to be able to catalyze the selective photo-induced oxidation of organic substrates in the presence of $[RuII(bpy)_3]^{2+}$ as a photosensitizer and $[CoIII(NH_3)_5Cl]^{2+}$ as a sacrificial electron acceptor, using water as an oxygen atom source.

The two published reviews describe different subjects of interest to the fields of biocatalysis and mix metallic-biocatalysis respectively.

The review by Anika Scholtissek et al. [8] describes the state-of-the-art of ene-reductases from the old yellow enzyme family (OYEs) to catalyze the asymmetric hydrogenation of activated alkenes to produce chiral products with industrial interest. The dependence of OYEs on pyridine nucleotide coenzyme can be avoided by using nicotinamide coenzyme mimetics. In the review, three main types of OYEs classification are described and characterized.

The review by Yajie Wang and Huimin Zhao [9] highlights some of the recent examples in the past three years that combined transition metal catalysis with enzymatic catalysis. With recent advances in protein engineering, catalyst synthesis, artificial metalloenzymes, and supramolecular assembly, there is great potential to develop more sophisticated tandem chemoenzymatic processes for the synthesis of structurally complex chemicals.

In conclusion, these nine publications give an overview of the possibilities of different catalysts, both traditional biocatalysts and hybrids with metals or organometallic complexes, to be used in different processes, in particular in synthetic reactions at very mild reaction conditions.

**Author Contributions:** Both authors contributed in writing the manuscript.

**Funding:** This work was supported by the Spanish Government (AGL2017-84614-C2-1-R and AGL2017-84614-C2-2-R).

**Conflicts of Interest:** The authors declare no conflict of interest.

## References

1. Vitale, P.; Digeo, A.; Perna, F.M.; Agrimi, G.; Salomone, A.; Scilimati, A.; Cosimo Cardellicchio, C.; Capriati, V. Stereoselective Chemoenzymatic Synthesis of Optically Active Aryl-Substituted Oxygen-Containing Heterocycles. *Catalysts* **2017**, *7*, 37. [CrossRef]

2. Wang, S.-Y.; Laborda, P.; Lu, A.-M.; Duan, X.-C.; Ma, H.-Y.; Liu, L.; Voglmeir, J. *N*-acetylglucosamine 2-Epimerase from *Pedobacter heparinus*: First Experimental Evidence of a Deprotonation/Reprotonation Mechanism. *Catalysts* **2016**, *6*, 212. [CrossRef]

3. Andrews, F.H.; Wechsler, C.; Rogers, M.P.; Meyer, D.; Tittmann, K.; McLeish, M.J. Mechanistic and Structural Insight to an Evolved Benzoylformate Decarboxylase with Enhanced Pyruvate Decarboxylase Activity. *Catalysts* **2016**, *6*, 190. [CrossRef]

4. Wang, X.; Yang, L.; Cao, W.; Ying, H.; Chen, K.; Ouyang, P. Efficient Production of Enantiopure D-Lysine from L-Lysine by a Two-Enzyme Cascade System. *Catalysts* **2016**, *6*, 168. [CrossRef]

5. Rivero, C.W.; Palomo, J.M. Covalent Immobilization of *Candida rugosa* Lipase at Alkaline pH and Their Application in the Regioselective Deprotection of Per-*O*-acetylated Thymidine. *Catalysts* **2016**, *6*, 115. [CrossRef]

6. Alnoch, R.C.; Rodrigues de Melo, R.; Palomo, J.M.; Maltempi de Souza, E.; Krieger, N.; Mateo, C. New Tailor-Made Alkyl-Aldehyde Bifunctional Supports for Lipase Immobilization. *Catalysts* **2016**, *6*, 191. [CrossRef]

7. Herrero, C.; Nguyen-Thi, N.; Hammerer, F.; Banse, F.; Gagné, D.; Doucet, N.; Mahy, J.-P.; Ricoux, R. Photoassisted Oxidation of Sulfides Catalyzed by Artificial Metalloenzymes Using Water as an Oxygen Source. *Catalysts* **2016**, *6*, 202. [CrossRef]

8. Scholtissek, A.; Tischler, D.; Westphal, A.H.; van Berkel, W.J.H.; Paul, C.E. Old Yellow Enzyme-Catalysed Asymmetric Hydrogenation: Linking Family Roots with Improved Catalysis. *Catalysts* **2017**, *7*, 130. [CrossRef]

9. Wang, Y.; Zhao, H. Tandem Reactions Combining Biocatalysts and Chemical Catalysts for Asymmetric Synthesis. *Catalysts* **2016**, *6*, 194. [CrossRef]

*catalysts*

MDPI

*Communication*

# Stereoselective Chemoenzymatic Synthesis of Optically Active Aryl-Substituted Oxygen-Containing Heterocycles [†]

**Paola Vitale [1,\*], Antonia Digeo [1], Filippo Maria Perna [1], Gennaro Agrimi [2,3], Antonio Salomone [4], Antonio Scilimati [1], Cosimo Cardellicchio [5] and Vito Capriati [1,\*]**

[1]  Dipartimento di Farmacia-Scienze del Farmaco, Università degli Studi di Bari «Aldo Moro», Consorzio C.I.N.M.P.I.S., Via E. Orabona 4, I-70125 Bari, Italy; anto41989@gmail.com (A.D.); filippo.perna@uniba.it (F.M.P.); antonio.scilimati@uniba.it (A.S.)

[2]  Department of Biosciences, Biotechnologies and Biopharmaceutics, University of Bari, Via E. Orabona 4, I-70125 Bari, Italy

[3]  Consorzio C.I.R.C.C. Via Celso Ulpiani 27, I-70126 Bari, Italy; gennaro.agrimi@uniba.it

[4]  Dipartimento di Scienze e Tecnologie Biologiche ed Ambientali, Università del Salento, Prov.le Lecce-Monteroni, I-73100 Lecce, Italy; antonio.salomone@unisalento.it

[5]  CNR ICCOM, Dipartimento di Chimica, Università di Bari, Via E. Orabona 4, I-70125 Bari, Italy; cardellicchio@ba.iccom.cnr.it

\*  Correspondence: paola.vitale@uniba.it (P.V.); vito.capriati@uniba.it (V.C.); Tel.: +39-0805442734 (P.V.); +39-0805442174 (V.C.); Fax: +39-0805442539 (P.V. & V.C.)

[†]  Dedicated to Professor Luigino Troisi on the occasion of his retirement.

Academic Editors: Jose M. Palomo and Cesar Mateo
Received: 22 December 2016; Accepted: 17 January 2017; Published: 25 January 2017

**Abstract:** A two-step stereoselective chemoenzymatic synthesis of optically active α-aryl-substituted oxygen heterocycles was developed, exploiting a whole-cell mediated asymmetric reduction of α-, β-, and γ-chloroalkyl arylketones followed by a stereospecific cyclization of the corresponding chlorohydrins into the target heterocycles. Among the various whole cells screened (baker's yeast, *Kluyveromyces marxianus* CBS 6556, *Saccharomyces cerevisiae* CBS 7336, *Lactobacillus reuteri* DSM 20016), baker's yeast was the one providing the best yields and the highest enantiomeric ratios (up to 95:5 er) in the bioreduction of the above ketones. The obtained optically active chlorohydrins could be almost quantitatively cyclized in a basic medium into the corresponding α-aryl-substituted cyclic ethers without any erosion of their enantiomeric integrity. In this respect, valuable, chiral non-racemic functionalized oxygen containing heterocycles (e.g., (*S*)-styrene oxide, (*S*)-2-phenyloxetane, (*S*)-2-phenyltetrahydrofuran), amenable to be further elaborated on, can be smoothly and successfully generated from their prochiral precursors.

**Keywords:** whole cell biocatalyst; baker's yeast; enantioselective bioreduction; oxiranes; oxetanes; tetrahydrofurans; halohydrins; chloroketones; oxygen-containing heterocycles; chemoenzymatic synthesis

## 1. Introduction

Oxygen-containing heterocycles are ubiquitous in natural products and biologically active compounds, and are also very common in many blockbuster pharmaceuticals [1,2]. The chemistry of saturated oxygen heterocycles is a topic of growing interest, and several papers dealing with more efficient methodologies for their preparation and their synthetic utility have been increasingly published. Epoxides, in particular, have been widely used in preparative chemistry [3,4] and in the asymmetric synthesis of fine chemicals and drugs (e.g., sertraline, nifenalol, Figure 1) [5–8] because of their versatility related to the ring strain. The oxetane skeleton is present in several natural organic products (e.g., oxetanocin, taxol, mitrophorone), and represents a versatile building block for the construction of biologically active compounds (e.g., EDO, Figure 1), or other valuable heterocyclic compounds [9–11]. It is also of interest in medicinal chemistry for the isosteric replacement of both the carbonyl and the *gem*-dimethyl group [12–15]. Asymmetric syntheses of optically active tetrahydrofurans have also been extensively investigated in the last few decades [16] because of their presence in many natural products and biologically active compounds (e.g., Goniothalesdiol, Figure 1). The preparation of chiral tetrahydrofurans has been efficiently performed by asymmetric cycloetherifications of hydroxy olefins in the presence of organocatalysts [17] or transition metals [18], or by the catalytic asymmetric hydrogenation of substituted furans [19].

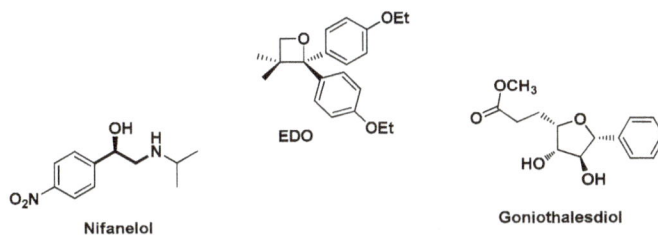

**Figure 1.** Drugs derived from optically active oxygen-containing heterocycles.

Optically active halohydrins have been successfully employed for the preparation of several chiral non-racemic oxygenated heterocycles (e.g., epoxides, oxetanes, tetrahydrofurans, pyrans).

Some general examples of stereoselective syntheses of halohydrins, as precursors of optically active cyclic ethers, are (a) the reduction of halogen-substituted ketones by means of hydrides complexed with chiral ligands (e.g., CBS-catalyst) [20,21]; (b) stereoselective hydrogenation processes run in the presence of Rh/Ru catalysts [22–24]; (c) microbial [25–28] or isolated enzymes-mediated [29] stereoselective reductions of α-halo-acetophenones; and (d) the kinetic resolution of racemic mixtures using dehalogenases (e.g., HheC from *Agrobacterium radiobacter AD1*) [30,31]. Our group recently focused on the development of new bio-catalyzed whole-cell biotransformations for the enantioselective preparation of chiral secondary alcohols, which are valuable precursor compounds for active pharmaceutic ingredients (APIs) [32–35].

Biocatalytic methodologies have received a great deal of attention for the asymmetric synthesis of biologically active molecules (also in industrial production) because of their high chemo-, regio-, and stereoselective performance under mild reaction conditions [36–38]. Building on these findings, herein we describe a chemoenzymatic synthetic strategy to prepare optically active epoxides, oxetanes, and tetrahydrofurans, which is based on the enantioselective bioreduction of α-, β-, and γ-haloketones in the presence of whole cell biocatalysts, followed by stereospecific cyclization of the corresponding enantio-enriched halohydrins (Scheme 1).

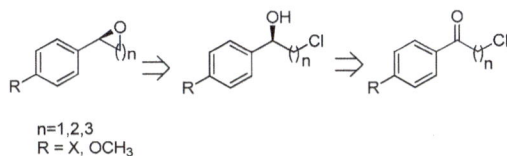

n=1,2,3
R = X, OCH₃

**Scheme 1.** A chemoenzymatic approach for the synthesis of optically active epoxides, oxetanes, and tetrahydrofurans via enantioselective bioreduction of halo-ketones with whole-cell biocatalysts.

## 2. Results

### 2.1. Screening of Biocatalysts for the Stereoselective Reduction of 3-Chloro-1-Arylpropanones

Various microorganisms are known to express different alcohol dehydrogenases (ADHs), each one exhibiting a specific stereo-preference according to the species, the metabolic growth conditions and phase, and the substrate specificity. To date, different yeasts have proven to be effective for the synthesis of functionalized styrene oxides with high stereoselectivity [39], whereas whole-cell biocatalysts with different stereo-preferences (e.g., *Kluyveromyces marxianus*, *Lactobacillus reuteri*) have been successfully employed for the preparation of enantio-enriched secondary alcohols [32–35]. With the aim of identifying the best whole-cell biocatalyst able to reduce different chloroketones with high enantioselectivity, we started our study by screening various biocatalysts for the stereoselective reduction of 3-chloro-1-arylpropanones (Table 1).

In the presence of 0.1 g/L resting cells (RC) of baker's yeast, chlorohydrin (*S*)-**2a** could be isolated with a 42% yield and in up to a 94:6 enantiomeric ratio (er) (Table 1, entry 1) starting from 3-chloro-1-phenylpropanone (**1a**), whereas the reduction in the presence of *Saccharomyces cerevisiae* CBS 7336 (GC) furnished (*S*)-**2a** with a 48% chemical yield and lower er (75:25) (Table 1, entry 2). In the presence of growing cells (GC) of *Kluyveromyces marxianus* CBS 6556, a mixture of products was detected in the reaction crude after 24 h incubation at 30 °C, and (*S*)-**2a** was isolated with only a 31% yield and almost in a racemic form (58:42 er) (Table 1, entry 3). The same biotransformation run in the presence of *Lactobacillus reuteri* DSM 20016 (RC) whole cells did not afford the desired chlorohydrin, with the main reaction being instead the dehydroalogenation of the starting haloketone and the formation of other minor products (see Supporting Information), as observed for other biocatalysts [40].

**Table 1.** Screening of biocatalysts for the stereoselective reduction of 3-chloro-1-aryl-propanones [a].

| Entry | Biocatalyst | Ar | Ketone 1 | Product 2 (Yield %) [b] | Conversion % | er [c] | Abs. Conf. [d] |
|---|---|---|---|---|---|---|---|
| 1 | *Baker's yeast* (RC) | $C_6H_5$ | **1a** | **2a** (42) | 50 | 94:6 | *S* |
| 2 | *Saccharomyces cerevisiae* (GC) [e] | $C_6H_5$ | **1a** | **2a** (48) | 55 | 75:25 | *S* |
| 3 | *Kluyveromyces marxianus* (GC) [f] | $C_6H_5$ | **1a** | **2a** (31) [g] | 70 | 58:42 | *S* |
| 4 | *Baker's yeast* (RC) | 4-F$C_6H_4$ | **1b** | **2b** (13) | 15 | 63:37 | *S* |
| 5 | *Baker's yeast* (RC) | 4-Br$C_6H_4$ | **1c** | **2c** (5) [h] | 85 | 95:5 | *S* |
| 6 | *Baker's yeast* (RC) | 4-MeO$C_6H_4$ | **1d** | **2d** (–) | 12 | ND [i] | ND [i] |

[a] Typical reaction conditions: orbital incubator (200 rpm); temperature: 30 °C; (GC): inoculum after 24 h growth in a sterile medium containing glucose (1%), peptone (0.5%), yeast extract (0.3%), and malt extract (0.3%) in sterile water; (RC): 0.1 g/L of cell wet mass in 0.1 M $KH_2PO_4$ buffer (pH = 7.4) enriched with 1% glucose, halo-ketone (2 mM final concentration); [b] Isolated yield after column chromatography; [c] Enantiomeric ratio (er) determined by HPLC analysis; [d] Absolute configuration (abs. conf.) of halohydrins (**2a–d**) determined by comparing optical rotation sign and retention time (HPLC analysis) with known data; [e] CBS 7536; [f] CBS 6556; [g] Propiophenone (35%) and 1-phenylpropan-1-ol (33%) have been detected by $^1$H NMR analysis of the reaction crude. [h] Propiophenone (75%) was isolated as the main product, together with 4-bromophenyloxetane (9%, er = 96:4); [i] ND means not determined because of the trace content.

Electronic effects of substituents present on the aromatic ring were also investigated. Upon reduction of 3-chloro-1-(4-fluorophenyl)-1-propanone (**1b**) with baker's yeast (RC), the corresponding alcohol (*S*)-**2b** was formed with a 13% yield and 63:37 er only (Table 1, entry 4), whereas *Lactobacillus reuteri* DSM 20016 (RC) was ineffective (see Table S1, Supporting Information). 1-(4-Bromophenyl)-3-chloro-1-propanone (**1c**) mainly underwent a dechlorination reaction with baker's yeast (RC), furnishing the corresponding propiophenone as the main product (75% yield) together with a small amount of 4-bromophenyloxetane (9% yield), though highly enantio-enriched (96:4 er).

The expected chlorohydrin (*S*)-**2c** formed with a 5% yield only but with 95:5 er (Table 1, entry 5). Finally, the action of baker's yeast (RC) on 1-(4-methoxyphenyl)-3-chloro-1-propanone (**1d**) produced the corresponding propiophenone (5%) as the result of a dehalogenation reaction of the starting ketone. Of note, such a dehalogenation reaction took place at 37 °C also in the absence of yeast. Thus, the elimination reaction was found to be independent from the biocatalyst [41,42], different from the behavior of the other microorganisms [43].

### 2.2. Screening of Biocatalysts for the Stereoselective Reduction of 4-Chloro-1-Aryl-1-Butanones

Several 4-chloro-1-aryl-1-butanones **1e–h** were also incubated and screened with various whole-cell biocatalysts. Baker's yeast (RC) mediated bioreduction of **1e** took place with moderate yield (44%), affording the chlorohydrin (*S*)-**2e** with an excellent 95:5 er (Table 2, entry 1).

**Table 2.** Screening of biocatalysts for the stereoselective reduction of 4-chloro-1-aryl-1-butanones [a].

| Entry | Biocatayst | Ar | Substrate 1 | Product 2 (Yield %) [b] | Conversion (%) | er [c] | Abs. Conf. [d] |
|---|---|---|---|---|---|---|---|
| 1 | *Baker's yeast* (RC) | $C_6H_5$ | **1e** | **2e** (44) | 49 | 95:5 | *S* |
| 2 | *S. cerevisiae* (GC) [e] | $C_6H_5$ | **1e** | **2e** (65) | 70 | 49:51 | *S* |
| 3 | *K. marxianus* (GC) [f] | $C_6H_5$ | **1e** | **2e** (4) | 7 | 42:58 | *S* |
| 4 | *Baker's yeast* (RC) | $4\text{-}FC_6H_4$ | **1f** | **2f** (–) [g] | 40 | ND [h] | ND [h] |
| 5 | *Baker's yeast* (RC) | $4\text{-}BrC_6H_4$ | **1g** | **2g** [i] | – [i] | ND [h] | ND [h] |
| 6 | *Baker's yeast* (RC) | $4\text{-}CH_3OC_6H_4$ | **1h** | **2h** (–) [j] | 5 | ND [h] | ND [h] |

[a] Typical reaction conditions: orbital incubator (200 rpm); temperature: 30 °C; (GC): inoculum after 24 h cell growth in a sterile medium containing glucose (1%), peptone (0.5%), yeast extract (0.3%), and malt extract (0.3%) in sterile water; (RC): 0.1 g/L of cell wet mass in 0.1 M $KH_2PO_4$ buffer (pH = 7.4) enriched with 1% glucose, haloketone (2 mM final concentration); [b] Isolated yield after column chromatography; [c] Enantiomeric ratio (er) determined by HPLC analysis; [d] Absolute configuration (abs. conf.) of halohydrins (**2e–h**) determined both by comparing optical rotation sign and retention time (HPLC analysis) with known data; [e] CBS 7336; [f] CBS 6556; [g] The corresponding butyrophenone (37%) has been detected by $^1$H NMR analysis of the reaction crude; [h] ND means not determined because of the trace content; [i] No reaction. [j] Chlorohydrin **2h** (5%) has been detected by GC-MS analysis of the reaction crude.

The yields increased up to 65% working with *Saccharomyces cerevisiae* CBS 7336 (GC), even if the corresponding halohydrin was isolated as a racemic mixture (49:51 er) (Table 2, entry 2). *Kluyveromyces marxianus* CBS 6556 (GC) reduced the halo-ketone **1e** both in low yield and enantioselectivity (Table 2, entry 3), whereas *Lactobacillus reuteri* DSM 20016 (RC) promoted the formation of 4-hydroxy-1-phenylbutanone as the only product, by the halogen substitution with a water molecule (Table S2, Supporting Information) [44]. As in the case of 1-arylpropanones, the baker's yeast performance was the best in terms of chemo- and stereo-selectivity. However, the reduction of different aryl-substituted γ-chloro-butyrophenones **1f–h** bearing electron-withdrawing and electron-donating groups proceeded sluggishly in water, presumably because of the poor solubility of the substrates in the used reaction medium or because of the lower intrinsic ketone reactivity, the main products being dehalogenated or hydroxy-substituted derivatives (Table 2 entries 4–6). Thus, the lower bioreduction reaction rates, corresponded to increasingly competitive dehalogenation reactions.

### 2.3. Screening of Biocatalysts for the Stereoselective Reduction of 2-Chloro-1-Acetophenones

The enantioselective reduction of functionalized α-haloacetophenones by baker's yeast is well-known [45], as well as the synthesis of optically active styrene oxides from haloketones by using isolated alcohol dehydrogenases (e.g., LkDHs from *Lactobacillus kefir*) [46]. Wild-type whole-cell biocatalysts are often preferred as biocatalysts over isolated and purified enzymes because they are cheaper than isolated and purified enzymes, easy to handle, and have a continuous source of enzymes and efficient internal cofactor (e.g., NAD(P)H) regeneration systems [39,47]. Building on our recent studies on the *anti*-Prelog stereo-preference of *Lactobacillus reuteri* DSM 20016 in the bioreduction of acetophenones [32], we investigated the possibility of preparing both the enantiomers of chiral aryl-epoxides **3i,j** (Table 4) carrying out the biotransformations in the presence of either baker's yeast or *Lactobacillus reuteri* DSM 20016 whole cells, followed by cyclization in a basic medium of the corresponding halohydrins **2i,2j** (Table 3).

**Table 3.** Screening of biocatalysts for the stereoselective reduction of 2-chloro-1-arylethanones.

| Entry | Biocatayst | Ar | Substrate | Chlorohydrin 2 (Yield %) [a] | Conversion (%) | Er [b] | Abs. Conf. [c] |
|-------|-----------|-----|-----------|------------------------------|----------------|--------|----------------|
| 1 | Baker's yeast [d] | $C_6H_5$ | **1i** | **2i** (53) | 55 | 90:10 | R |
| 2 | Baker's yeast | $4\text{-}ClC_6H_4$ | **1j** | **2j** (64) | 70 | 63:37 | R |
| 3 | L. reuteri (RC) [e] | $4\text{-}ClC_6H_4$ | **1j** | **2j** (28) | 30 | 96:4 | S |

[a] Isolated yield after column chromatography; [b] Enantiomeric ratio (er) determined by HPLC analysis; [c] Absolute configuration (abs. conf.) of halohydrins determined by comparing optical rotation sign with known data; [d] Typical reaction conditions: orbital incubator: 200 rpm; temperature: 30 °C; haloketone (2 mM final concentration) was added to a 0.1 g/L of cell wet mass suspended in tap water (RC); [e] Typical reaction conditions: cells were suspended in PBS at pH 7.4 supplemented with 1% glucose; then, ketone was added at the final concentration of 1 g/L (50 mL total volume), anaerobiosis; temperature: 37 °C; orbital incubator: 200 rpm; [e] DSM 20016.

Baker's yeast successfully reduced α-chloroacetophenone **1i** and α-chloro-*p*-chloroacetophenone **1j** providing the expected chlorohydrins (*R*)-**2i** and (*R*)-**2j** with 53% and 64% yields, respectively, and with up to 90:10 er after 24 h incubation at 30 °C (Table 3, entries 1, 2). On the other hand, the *anti*-Prelog stereo-preference of *Lactobacillus reuteri* DSM 20016 [10,32] furnished (*S*)-**2j** with a 28% yield but with a higher stereoselectivity (96:4 er) in comparison with baker's yeast (Table 3, entry 3). Thus, baker's yeast and *Lactobacillus reuteri* DSM 20016 behave as two complementary whole cell biocatalysts for the synthesis of optically active 2-chloro-1-arylethanols because of their ADHs opposite stereo-preference, though with their own substrate specificity (Table 3, entries 2, 3).

### 2.4. Synthesis of Optically Active 2-Aryloxetanes, 2-Phenyltetrahydrofurans, 2-Arylepoxides

Stereospecific cyclization in basic conditions (*t*-BuOK/THF or NaOH/*i*PrOH, room temperature) of enantio-enriched chlorohydrins **2a**, **2c**, **2e**, **2i**, and **2j** obtained from baker's yeast (*vide supra*) took place smoothly, providing almost quantitatively the corresponding (*S*)-2-aryloxetanes **3a,c**, (*S*)-2-phenyltetrahydrofuran (**3e**), and (*S*)-styrene oxide (**3i**) with high er (up to 96:4) (Table 4, entries 1–4). On the other hand, (*R*)-*p*-chlorostyrene oxide **3j** was isolated with a 97% yield and with er = 96:4 further to the bioreduction of **1j** with *L. reuteri* DSM 20016 (Table 4, entry 5). Thus, two terminal enantiomeric arylepoxides could be synthesized exploiting the opposite stereo-preference of two cheap and complementary biocatalysts.

**Table 4.** Synthesis of optically active 2-aryloxygenated heterocycles **3** from halohydrins **2** [a].

| Entry | Ar | Chlorohydrin 2 (er) | n | Product 3 (Yield %) [b] | er [c] | Abs. Conf. [d] |
|-------|------|---------------------|------|-------------------------|--------|----------------|
| 1 | $C_6H_5$ | (*S*)-**2a** (94:6) | 2 | **3a** (98) | 95:5 | *S* |
| 2 | 4-BrC$_6$H$_4$ | (*S*)-**2c** (95:5) | 2 | **3c** (98) | 96:4 | *S* |
| 3 | $C_6H_5$ | (*S*)-**2e** (95:5) | 3 | **3e** (98) | 95:5 | *S* |
| 4 | $C_6H_5$ | (*R*)-**2i** (90:10) | 1 [e] | **3i** (95) | 90:10 | *R* |
| 5 | 4-ClC$_6$H$_4$ | (*S*)-**2j** (96:4) | 1 [e] | **3j** (97) | 96:4 | *S* |

[a] Typical reaction conditions: chlorohydrin **2** (1 mmol), *t*-BuOK (3 mmol), THF (5 mL), 25 °C, 4 h; [b] Isolated yield after column chromatography; [c] Enantiomeric ratio (er) determined by GC analysis; [d] Absolute configuration (abs. conf.) of cyclic ethers **3** determined by comparing optical rotation sign with known data; [e] NaOH (3 mL, 1 N) as the base and *i*-PrOH (2 mL) as the solvent were used instead of *t*-BuOK and THF.

## 3. Materials and Method

### 3.1. General Methods

[1]H NMR and [13]C NMR spectra were recorded on a Bruker Avance 600 MHz (Bruker, Milan, Italy) or Varian Inova 400 MHz spectrometer (Agilent Technologies, Santa Clara, CA, USA) and chemical shifts are reported in parts per million ($\delta$).[19]F NMR spectra were recorded by using CFCl$_3$ as an internal standard. Absolute values of the coupling constants are reported. FT-IR spectra were recorded on a Perkin-Elmer 681 spectrometer (Perkin Elmer, Waltham, MA, USA). GC analyses were performed on a HP 6890 model Series II (Agilent Technologies, Santa Clara, CA, USA) by using a HP1 column (methyl siloxane; 30 m × 0.32 mm × 0.25 μm film thickness). Thin-layer chromatography (TLC) was carried out on pre-coated 0.25 mm thick plates of Kieselgel 60 F$_{254}$; visualisation was accomplished by UV light (254 nm) or by spraying a solution of 5% (*w/v*) ammonium molybdate and 0.2% (*w/v*) cerium(III) sulfate in 100 mL 17.6% (*w/v*) aq. sulfuric acid and heating to 200 °C until blue spots appeared. Column chromatography was conducted by using silica gel 60 with a particle size distribution of 40–63 μm and 230–400 ASTM. Petroleum ether refers to the 40–60 °C boiling fraction. GC-MS analyses were performed on a HP 5995C model (Agilent Technologies, Santa Clara, CA, USA) and elemental analyses on an Elemental Analyzer 1106-Carlo Erba-instrument (Carlo-Erba, Milan, Italy). MS-ESI analyses were performed on an Agilent 1100 LC/MSD trap system VL (Agilent Technologies, Santa Clara, CA, USA). Optical rotation values were measured at 25 °C using a Perkin Elmer 341 polarimeter (Perkin Elmer, Waltham, MA, USA) with a cell of 1 dm path length; the concentration (*c*) is expressed in g/100 mL. The enantiomeric ratios were determined by HPLC analysis using an Agilent 1100 chromatograph (Agilent Technologies, Waldbronn, Germany), equipped with a DAD detector, and Phenomenex LUX Cellulose-1 [Cellulose tris(3,5-dimethylphenylcarbamate)], LUX Cellulose-2 [Cellulose 2 tris(3-chloro-4-methylphenylcarbamate)], and LUX Cellulose-4 [Cellulose tris(4-chloro-3-methylphenylcarbamate)] columns (250 × 4.6 mm), or by GC-analyses performed on a Hewlett–Packard 6890 Series II chromatograph (Agilent Technologies, Inc., Wilmington, DE, USA) equipped with a Chirasil-DEX CB (250 × 0.25 μm) capillary column, column head pressure = 18 psi, He flow 2 mL/min, split ratio 100/1, *T* (oven) from 90 to 120 °C. All the chemicals and

solvents were of commercial grade and were further purified by distillation or crystallization prior to use. All optically active halohydrins **2a–j** and oxygen-containing heterocycles **3a–j** obtained by bioreductions of halo-ketones had analytical and spectroscopic data identical to those previously reported or to the commercially available compounds. Racemic mixtures (for HPLC references) were synthesized by $NaBH_4$ reduction in EtOH with 87%–96% yields according to the reported procedures [32–35].

### 3.2. Microorganism and Cultures

*Saccharomyces cerevisiae* CBS 7336 and *Kluyveromyces marxianus* CBS 6556 were obtained from public type culture collections (CBS, DSM, Delft, The Netherlands) under aerobic conditions in a medium containing 0.3% yeast extract, 0.3% malt extract, 0.5% peptone, and 1% glucose. Agar-agar (2%) was added to the same medium for cell preservation on agar slants.

*Lactobacillus reuteri* DSM 20016 was obtained from a DSMZ culture collection (Braunschweig, Germany) [48]. Cells were maintained at −80 °C in culture broth supplemented with 25% (w/v) glycerol. Pre-cultures and cultures were carried out in a classical MRS medium [49] (Oxoid) containing 20 g/L glucose, 10 g/L peptone, 8 g/L meat extract, 4 g/L yeast extract, 1 g/L Tween 80, 2 g/L di-potassium hydrogen phosphate, 5 g/L sodium acetate·$3H_2O$, 2 g/L tri-ammonium citrate, 0.2 g/L of magnesium sulfate·$7H_2O$, and 0.05 g/L manganese sulfate·$2H_2O$. Cells were incubated at 37 °C for 24 h, statically. Cell density was monitored using optical density at 620 nm ($OD_{620}$) with a Genesys TM 20 spectrophotometer (Thermo Fisher Scientific Inc., Waltham, MA, USA).

### 3.3. Blank Experiments

A 1 L flask containing 400 mL of the culture medium was stirred at 30 °C on an orbital shaker at 200 rpm. Halo-ketones **1a–j** (50 mg) were added. The reaction was monitored by TLC and stopped after 24 h. The content of the flask was extracted with $Et_2O$ and analyzed by GC-MS or $^1H$ NMR analysis.

### 3.4. Bioreduction of Haloketones 1a,e by Yeasts Growing Cells: General Procedure

Cells preserved on agar slants at 4 °C were used to inoculate 250 mL flasks containing 100 mL of the culture medium. The flasks were incubated aerobically at 30 °C on an orbital shaker and stirred at 250 rpm. Flasks (250 mL) containing 100 mL of the culture medium were then inoculated with 5 mL of the 24-h-old suspension and incubated in the same conditions for 24 h. Flasks (1 L) containing 400 mL of the culture medium were then inoculated with 5 mL of the latter suspension and incubated for 24 h. The optical density was checked at 620 nm for all cultures before adding halo-ketones **1a,e** (100 mg) previously dissolved in 1 mL of EtOH. The progress of the reactions was monitored by TLC and/or GC and stopped after 24 h, as indicated in Tables 1 and 2. The content of the flask was then centrifuged and the supernatant extracted with EtOAc. All the reactions were repeated at least twice without any detectable bias in the results. Silica gel column chromatography of the reaction crude, using hexane and EtOAc (90:10 or 80:20) as the eluents yielded the desired halohydrins (**2a,e**) (Tables 1 and 2).

### 3.5. Baker's Yeast Bioreductions General Procedure

Baker's yeast (15 g) was dispersed to give a smooth paste in tap water (250 mL). The substrate (100 mg) was added and stirred at 30 °C in an orbital shaker (200 rpm). The reaction progress was monitored by TLC. After 24 h (Tables 1–3), the reaction was stopped by centrifugation, decantation, and extraction by EtOAc or $CH_2Cl_2$. The extracts were dried over anhydrous $Na_2SO_4$ and the solvent was evaporated under reduced pressure. The residue was purified by silica gel column chromatography using hexane and EtOAc (10:1 or 8:2) as the eluents to yield the desired halohydrins (**2a–j**) (Tables 1–3).

## 3.6. Characterization Data of Compounds 2a-c,e,i,j and 3a,c,e,i,j

(*S*)-3-Chloro-1-phenylpropan-1-ol (**2a**) [50]. 42% Yield (from baker's yeast), $R_f$ 0.50 (2:8 ethyl acetate:hexane); $[\alpha]_D^{20}$ = −16.9 (*c* 1, CHCl$_3$), er [*S*]:[*R*] = 94:6 determined by HPLC [LUX Cellulose-1 column (hexane:2-propanol = 90:10), 0.8 mL/min], $t_R$ [major (*S*)-enantiomer] = 11.9 min; $t_R$ [minor (*R*)-enantiomer] = 12.8 min. $^1$H NMR (CDCl$_3$, 600 MHz, 25 °C, δ): 7.38–7.25 (m, 5 H, aromatic protons), 4.97–4.95 (m, 1 H, C*H*OH), 3.77–3.73 (m, 1 H, CH*H*Cl), 3.59–3.55 (m, 1 H, C*H*HCl), 2.28–2.22 (m, 1 H, C*H*H), 2.13–2.08 (m, 1 H, CH*H*), 1.97–1.89 (bs, 1 H, O*H*, exchanges with D$_2$O). $^{13}$C NMR (150 MHz, CDCl$_3$, 25 °C; δ): 143.7, 128.7, 127.9, 125.8, 71.3, 41.7, 41.4. GC-MS (70 eV) *m/z* (rel.int.): 172 [(*M* + 2)$^+$, 1], 170 (*M*$^+$, 3), 117(2), 115(2), 108(8), 107(100), 105(9), 79(49), 77(28), 31(8).

(*S*)-3-Chloro-1-(4′-fluorophenyl)propan-1-ol (**2b**) [51,52]. 13% Yield (from baker's yeast), $R_f$ 0.40 (1:15 ethyl acetate:hexane); $[\alpha]_D^{20}$ = −8.13° (*c* 0.75, CHCl$_3$), er [*S*]:[*R*] = 63:37, determined by HPLC [LUX Cellulose-1 column (hexane:2-propanol 90:10), 0.8 mL/min], $t_R$ [major (*S*)-enantiomer] = 9.8 min; $t_R$ [minor (*R*)-enantiomer] = 10.5 min. $^1$H NMR (CDCl$_3$, 400 MHz, 25 °C, δ): 7.36–7.33 (m, 2 H, aromatic protons), 7.07–7.03 (m, 2 H, aromatic protons), 4.96–4.93 (m, 1 H, C*H*OH), 3.77–3.70 (m, 1 H, CH*H*Cl), 3.57–3.52 (m, 1 H, C*H*HCl), 2.25–2.19 (m, 1 H, C*H*H), 2.10–2.03 (m, 1 H, CH*H*), 1.99–1.85 (bs, 1 H, O*H*, exchanges with D$_2$O). $^{13}$C NMR (CDCl$_3$, 125 MHz, 25 °C, δ): 41.5, 41.6, 70.7, 115.5 (d, $^2J_{C-F}$ = 21.0 Hz), 127.4 (d, $^3J_{C-F}$ = 8.0 Hz), 139.4, 162.3 (d, $^1J_{C-F}$ = 246.0 Hz). $^{19}$F NMR (376 MHz, CDCl$_3$, δ): −114.53, (m). GC-MS (70 eV) *m/z* (rel.int.): 188 (*M*$^+$, 3), 126(8), 125(100), 123(11), 97(46), 96(7), 95(15), 77(14).

(*S*)-3-Chloro-1-(4′-bromophenyl)propan-1-ol (**2c**) [53,54]. 5% Yield (from baker's yeast), $R_f$ 0.3 (1:10 ethyl acetate:hexane); $[\alpha]_D^{20}$ = −4.95° (*c* 0.75, CHCl$_3$), er [*S*]:[*R*] = 95:5, determined by GC with isotherm at 170 °C, $t_R$ [minor (*R*)-enantiomer] = 43.0 min; $t_R$ [major (*S*)-enantiomer] = 44.3 min. $^1$H NMR (CDCl$_3$, 400 MHz, 25 °C, δ): 7.49–7.45 (m, 2 H, aromatic protons), 7.25–7.22 (m, 2 H, aromatic protons), 4.14–4.08 (m, 1 H, C*H*OH), 3.77–3.70 (m, 1 H, CH*H*Cl), 3.57–3.52 (m, 1 H, C*H*HCl), 2.22–2.15 (m, 1 H, C*H*H), 2.05–2.00 (m, 1 H, CH*H*), 2.00–1.85 (bs, 1 H, O*H*, exchanges with D$_2$O). $^{13}$C NMR (CDCl$_3$, 150 MHz, 25 °C, δ): 142.7, 131.8, 131.7, 127.5, 121.7, 70.6, 41.5. GC-MS (70 eV) *m/z* (rel.int.): [(*M* + 4)$^+$, 2]; [(*M*+2)$^+$, 8]; (*M*$^+$, 6), 188 (7), 187 (91), 185 (100), 183 (5), 159 (13); 157 (17), 155 (5), 78 (21), 76 (5), 75 (5), 51 (8), 50 (5).

(*S*)-4-Chloro-1-phenylbutan-1-ol (**2e**) [55]: 44% Yield (from baker's yeast), $R_f$ 0.3 (1:10 ethyl acetate: hexane); $[\alpha]_D^{20}$ = −26° (*c* 1, CHCl$_3$), er [*S*]:[*R*] = 95:5, determined by HPLC [LUX Cellulose-1 coloumn (hexane:2-propanol = 90:10), 0.5 mL/min], $t_R$ [minor (*R*)-enantiomer] = 24.2 min; $t_R$ [major (*R*)-enantiomer] = 25.7 min. $^1$H NMR (CDCl$_3$, 400 MHz, 25 °C, δ): 7.40–7.27 (m, 5 H, aromatic protons), 4.74–4.71 (m, 1 H, C*H*OH), 3.60–3.53 (m, 2 H), 1.97–1.78 (m, 4 H), 1.85–1.80 (bs, 1 H, O*H*, exchanges with D$_2$O). $^{13}$C NMR (CDCl$_3$, 100 MHz, 25 °C, δ): δ 29.1, 36.4, 45.2, 74.0, 125.9, 128.0, 128.7, 144.5. GC-MS (70 eV) *m/z* (rel.int.): 186 [(*M*+2)$^+$, 0.4], 184 (*M*$^+$, 3), 126 (8), 108 (6), 107 (100), 105 (17), 91 (4), 79 (42), 78 (6), 77 (28).

(*R*)-2-Chloro-1-phenylethanol (**2i**): 53% yield, $R_f$ 0.4 (1:10 ethyl acetate:hexane); $[\alpha]_D^{20}$ = −40° (*c* 0.50, CHCl$_3$) from baker's yeast, er [*S*]:[*R*] = 90:10, determined by HPLC [LUX Cellulose-1 coloumn (hexane:2-propanol 90:10), 0.8 mL/min], $t_R$ [minor (*S*)-enantiomer] = 13.7 min; $t_R$ [major (*R*)-enantiomer] = 15.4 min. $^1$H NMR (CDCl$_3$, 600 MHz, 25 °C, δ): 7.41–7.38 (m, 5 H, aromatic protons), 4.92–4.91 (m, 1 H, C*H*OH), 3.77–3.75 (m, 1 H, CH*H*Cl), 3.68–3.65 (m, 1 H, C*H*HCl), 2.20 (bs, 1 H, O*H*, exchanges with D$_2$O).

(*S*)-2-Chloro-1-(4′-chlorophenyl)ethanol (**2j**): 28% Yield, $R_f$ 0.4 (2:8 ethyl acetate:hexane); $[\alpha]_D^{20}$ = +29° (*c* 0.3, CHCl$_3$) from *Lactobacillus reuteri*. er [*S*]:[*R*] = 96:4, determined by HPLC $t_R$ [minor (*R*)-enantiomer] = 17.4 min; $t_R$ [major (*S*)-enantiomer] = 17.8 min. $^1$H NMR (CDCl$_3$, 600 MHz, 25 °C, δ): 7.36–7.32 (m, 2 H), 7.20–7.16 (m, 2 H), 4.89–4.87 (m, 1 H, C*H*OH), 3.72–3.70 (m, 1 H, CH*H*Cl), 3.62–3.59 (m, 1 H, C*H*HCl), 3.30–2.60 (bs, 1 H, O*H*, exchanges with D$_2$O). GC-MS (70 eV) *m/z* (rel.int.): 192[(*M* + 2)$^+$, 3], 190 (*M*$^+$, 5), 143 (32), 142 (8), 141 (100), 113 (14), 78 (6), 77 (55); 51 (8), 50 (5), 49 (3).

(*S*)-2-Phenyloxetane (**3a**) [56–58]: 98% Yield, $[\alpha]_D^{20} = -4.3°$ (*c* 1, CHCl₃), er [S]:[R] = 95:5 determined by GC with isotherm at 90 °C, $t_R$ [major (*S*)-enantiomer] = 38.3 min; $t_R$ [minor (*R*)-enantiomer] = 40.8 min. ¹H NMR (CDCl₃, 400 MHz, 25 °C, δ): 7.48–7.27 (m, 5 H, aromatic protons), 5.69–5.65 (m, 1 H), 4.86–4.81 (m, 1 H), 4.69–4.50 (m, 1 H, C*H*H), 3.07–2.98 (m, 1 H, C*HH*), 2.72–2.63 (m, 1 H, CH*H*).

(*S*)-2-(4-Bromophenyl)oxetane (**3c**) [59]: 9% yield, er [S]:[R] = 96:4, determined by GC with isotherm at 150 °C, $t_R$ [minor (*R*)-enantiomer] = 22.5 min; $t_R$ [major (*S*)-enantiomer] = 23.1 min. ¹H NMR (CDCl₃, 600 MHz, 25 °C, δ): 7.47–7.45 (m, 2 H, aromatic protons), 7.21–7.10 (m, 2 H, aromatic protons), 5.69–5.59 (m, 1 H), 4.83–4.78 (m, 1 H), 4.69–4.63 (m, 1 H), 2.93–2.50 (m, 2 H.

(*S*)-2-Phenyltetrahydrofuran (**3e**) [60]: 98% yield, $[\alpha]_D^{20} = -1.6°$ (*c* 0.50, CHCl₃), er [S]:[R] = 95:5 determined by GC with isotherm at 110 °C, $t_R$ [major (*S*)-enantiomer] = 21.8 min; $t_R$ [minor (*R*)-enantiomer] = 22.9 min. ¹H NMR (CDCl₃, 400 MHz, 25 °C, δ): 7.38–7.26 (m, 5 H, aromatic protons), 4.74–4.71 (m, 1 H), 4.60–3.53 (m, 2 H), 1.98–1.79 (m, 4 H).¹³C NMR (CDCl₃, 150 MHz, 25 °C, δ): 144.3, 128.6, 127.8, 125.8, 73.9, 44.9, 36.2, 28.9.

(*R*)-Styrene oxide (**3i**) [61]: 95% yield, $[\alpha]_D^{20} = -25°$ (*c* 1, CHCl₃), er [R]:[S] = 90:10, determined by GC with isotherm at 100 °C, $t_R$ [major (*R*)-enantiomer] = 11.7 min; $t_R$ [minor (*S*)-enantiomer] = 12.3 min. ¹H NMR (CDCl₃, 600 MHz, 25 °C, δ): 7.31–7.22 (m, 5 H, aromatic protons), 3.83–81 (m, 1 H), 3.13–3.10 (m, 1 H); 2.91–2.87 (m, 1 H). ¹³C NMR (CDCl₃, 150 MHz, 25 °C, δ): 51.0, 52.2, 125.4, 128.1, 128.4, 137.5.

(*S*)-4-Chlorostyrene oxide (**3j**) [62]: 97% yield, $[\alpha]_D^{20} = +23°$ (*c* 1, CHCl₃), er [R]:[S] = 4:96 determined by GC with isotherm at 100 °C, $t_R$ [minor (*R*)-enantiomer] = 37.5 min; $t_R$ [major (*S*)-enantiomer] = 39.2 min. ¹H NMR (CDCl₃, 600 MHz, 25 °C, δ): 7.30–7.27 (m, 2 H, aromatic protons), 7.19–7.17 (m, 2 H, aromatic protons), 3.81–3.79 (m, 1 H), 3.12–3.10 (m, 1 H), 2.73–2.71 (m, 1 H). ¹³C NMR (CDCl₃, 150 MHz, 25 °C, δ): 51.3, 51.8, 126.8, 128.7, 133:9, 136.2.

## 4. Conclusions

In summary, stereo-defined aryl-substituted oxygen-containing heterocycles have been, for the first time, synthesized via a new chemoenzymatic approach based on the stereoselective whole-cell bioreduction of α-, β-, and γ-chloroalkyl arylketones into the corresponding chlorohydrins, followed by a final stereospecific cyclization. Among the different microorganisms screened (baker's yeast, *Kluyveromyces marxianus* CBS 6556, *Saccharomyces cerevisiae* CBS 7336, *Lactobacillus reuteri* DSM 20016) baker's yeast was the most efficient in providing chlorohydrins with the best isolated yields ranging from 42% to 64% and the highest er up to 95:5. 3-Chloropropiophenone, 4-chlorobutyrophenone, 4-chloro-4′-bromopropiophenone, and 2-chloroacetophenone have been reduced with good to moderate enantioselectivities by baker's yeast, whereas *Lactobacillus reuteri* DSM 20016 proved to be the best microorganism in performing the bioreduction of 2-chloro-4′-chloroacetophenone with an (*S*) absolute configuration and er up to 96:4. All the optically active chlorohydrins were subsequently stereo-specifically and almost quantitatively converted into optically active *S*-configured 2-aryloxetanes, 2-phenyltetrahydrofuran, and 2-arylepoxides without any erosion of the starting er. (*R*)-*p*-Chlorostyrene oxide could be prepared with the opposite configuration and in up to 96:4 er compared to baker's yeast, by subjecting to cyclization the α-chlorohydrin obtained from *Lactobacillus reuteri* DSM 20016. Since the wild-type whole-cell biocatalysts selected (baker's yeast and *Lactobacillus reuteri* DSM 20016) are cheap and commercially available, this methodology is auspicious for setting up industrially relevant and cost-effective biotransformations for a large-scale production of oxygen-containing heterocycles, and thus for the stereo-selective preparation of chiral drugs [18]. It is noteworthy that the tested substrates were slightly soluble in the aqueous solvents used in the above-mentioned biotransformations. Hence it is very likely that the yield can be further increased by simple process engineering approaches such as the fed-batch supply of the substrate or the use of bioreactors with carefully controlled operational conditions.

**Supplementary Materials:** The following are available online at www.mdpi.com/2073-4344/7/2/37/s1, Table S1: Screening of biocatalysts for the stereoselective reduction of 3-chloro-1-aryl-propanones, Table S2: Screening of biocatalysts for the stereoselective reduction of 4-chloro-1-aryl-1-butanones.

**Acknowledgments:** This work was financially supported by the University of Bari within the framework of the Project "Sviluppo di nuove metodologie di sintesi mediante l'impiego di biocatalizzatori e solventi a basso impatto ambientale" (code: Perna01333214Ricat), and by both the C.I.N.M.P.I.S. (Consorzio Interuniversitario Nazionale di Ricerca in Metodologie e Processi Innovativi di Sintesi) and C.I.R.C.C. (Interuniversity Consortium Chemical Reactivity and Catalysis) consortia. This work was also partially supported by the "Reti di Laboratori–Produzione Integrata di Energia da Fonti Rinnovabili nel Sistema Agroindustriale Regionale" program funded by the "Apulia Region Project Code 01". (Intervento cofinanziato dall'Accordo di Programma Quadro in materia di Ricerca Scientifica–II Atto Integrativo–PO FESR 2007–2013, Asse I, Linea 1.2-PO FSE 2007–2013 Asse IV "Investiamo nel vostro futuro".

**Author Contributions:** A.D. and P.V. conceived and designed the experiments; A.D. and G.A. performed the experiments; P.V., G.A., C.C., F.M.P., and A.S. analyzed the data; V.C., A.S., F.M.P., and C.C. contributed reagents/materials/analysis tools; V.C. and P.V. wrote the paper.

**Conflicts of Interest:** The authors declare no conflict of interest.

# References

1. Perna, F.M.; Salomone, A.; Capriati, V. Recent Developments in the Lithiation Reactions of Oxygen Heterocycles. In *Advances in Heterocyclic Chemistry*; Scriven, E.F.V., Ramsden, C.A., Eds.; Academic Press Inc.: Oxford, UK, 2016; Volume 118, pp. 91–127.

2. Capriati, V.; Perna, F.M.; Salomone, A. "The Great Beauty" of organolithium chemistry: a land still worth exploring. *Dalton Trans.* **2014**, *43*, 14204–14210. [CrossRef] [PubMed]

3. Florio, S.; Perna, F.M.; Salomone, A.; Vitale, P. Reduction of Epoxides. In *Comprehensive Organic Synthesis*, 2nd ed.; Molander, G.A., Knochel, P., Eds.; Elsevier: Oxford, UK, 2014; Volume 8, pp. 1086–1122.

4. Salomone, A.; Perna, F.M.; Sassone, F.C.; Falcicchio, A.; Bezenšek, J.; Svete, J.; Stanovnik, B.; Florio, S.; Capriati, V. Preparation of Polysubstituted Isochromanes by Addition of ortho-Lithiated Aryloxiranes to Enaminones. *J. Org. Chem.* **2013**, *78*, 11059–11065. [CrossRef] [PubMed]

5. Bayston, D.J.; Travers, C.B.; Polywka, M.E.C. Synthesis and evaluation of a chiral heterogeneous transfer hydrogenation catalyst. *Tetrahedron Asymmetry* **1998**, *9*, 2015–2018. [CrossRef]

6. Cross, D.J.; Kenny, J.A.; Houson, I.; Campbell, L.; Walsgrove, T.; Wills, M. Rhodium versus ruthenium: Contrasting behaviour in the asymmetric transfer hydrogenation of α-substituted acetophenones. *Tetrahedron Asymmetry* **2001**, *12*, 1801–1806. [CrossRef]

7. Morris, D.J.; Hayes, A.M.; Wills, M. The "Reverse-Tethered" Ruthenium (II) Catalyst for Asymmetric Transfer Hydrogenation: Further Applications. *J. Org. Chem.* **2006**, *71*, 7035–7044. [CrossRef]

8. Perrone, M.G.; Santandrea, E.; Giorgio, E.; Bleve, L.; Scilimati, A.; Tortorella, P. A chemoenzymatic scalable route to optically active (*R*)-1-(pyridin-3-yl)-2-aminoethanol, valuable moiety of β3-adrenergic receptor agonists. *Bioorg. Med. Chem.* **2006**, *14*, 1207–1214. [CrossRef] [PubMed]

9. Soai, K.; Niwa, S.; Yamanoi, T.; Hikima, H.; Ishizaki, M. Asymmetric synthesis of 2-aryl substituted oxetanes by enantioselective reduction of β-halogenoketones using lithium borohydride modified with *N,N'*-dibenzoylcystine. *J. Chem. Soc., Chem. Commun.* **1986**, 1018–1019. [CrossRef]

10. Vitale, P.; Perna, F.M.; Agrimi, G.; Scilimati, A.; Salomone, A.; Cardellicchio, C.; Capriati, V. Asymmetric chemoenzymatic synthesis of 1,3-diols and 2,4-disubstituted aryloxetanes by using whole cell biocatalysts. *Org. Biomol. Chem.* **2016**, *14*, 11438–11445. [CrossRef] [PubMed]

11. Lo, M.M.; Fu, G.C. Applications of Planar-Chiral Heterocycles in Enantioselective Catalysis: Cu(I)/Bisazaferrocene-Catalyzed Asymmetric Ring Expansion of Oxetanes to Tetrahydrofurans. *Tetrahedron* **2001**, *57*, 2621–2634. [CrossRef]

12. Malapit, C.A.; Howell, A.R. Pt-Catalyzed Rearrangement of Oxaspirohexanes to 3-Methylenetetrahydrofurans: Scope and Mechanism. *J. Org. Chem.* **2015**, *80*, 8489–8495. [CrossRef] [PubMed]

13. Shimada, H.; Hasegawa, S.; Harada, T.; Tomisawa, T.; Fujii, A.; Takita, T. Oxetanocin, a novel nucleoside from bacteria. *J. Antibiot.* **1986**, *39*, 1623–1625. [CrossRef] [PubMed]

14. Wani, M.C.; Taylor, H.L.; Wall, M.E.; Caggon, P.; McPhall, A.T. Plant antitumor agents. VI. The isolation and structure of taxol, a novel antileukemic and antitumor agent from Taxus brevifolia. *J. Am. Chem. Soc.* **1971**, *93*, 2325–2327. [CrossRef] [PubMed]

15. Wuitschik, G.; Carreira, E.M.; Wagner, B.; Fischer, H.; Parrilla, I.; Schuler, F.; Rogers-Evans, M.; Müller, K. Oxetanes in Drug Discovery: Structural and Synthetic Insights. *J. Med. Chem.* **2010**, *53*, 3227–3246. [CrossRef] [PubMed]

16. Jalce, G.; Franck, X.; Figadère, B. Diastereoselective synthesis of 2,5-disubstituted tetrahydrofurans. *Tetrahedron Asymmetry* **2009**, *20*, 2537–2650. [CrossRef]

17. Asano, K.; Matsubara, S. Asymmetric Catalytic Cycloetherification Mediated by Bifunctional Organocatalysts. *J. Am. Chem. Soc.* **2011**, *133*, 16711–16713. [CrossRef] [PubMed]

18. Murayama, H.; Nagao, K.; Ohmiya, H.; Sawamura, M. Copper(I)-Catalyzed Intramolecular Hydroalkoxylation of Unactivated Alkenes. *Org. Lett.* **2015**, *17*, 2039–2041. [CrossRef] [PubMed]

19. Pauli, L.; Tannert, R.; Scheil, R.; Pfaltz, A. Asymmetric Hydrogenation of Furans and Benzofurans with Iridium-Pyridine-Phosphinite Catalysts. *Chem. Eur. J.* **2015**, *21*, 1482–1487. [CrossRef] [PubMed]

20. Corey, E.J.; Shibata, S.; Bakshi, R.K. An efficient and catalytically enantioselective route to (*S*)-(-)-phenyloxirane. *J. Org. Chem.* **1988**, *53*, 2861–2863. [CrossRef]

21. Cordes, D.B.; Kwong, T.J.; Morgan, K.A.; Singaram, B. Chiral Styrene Oxides from α-haloacetophenones Using NaBH$_4$ and TarB-NO$_2$, a Chiral Lewis Acid. *Tetrahedron Lett.* **2006**, *47*, 349–351. [CrossRef]

22. Hamada, T.; Torii, T.; Izawa, K.; Noyori, R.; Ikariya, T. Practical Synthesis of Optically Active Styrene Oxides via Reductive Transformation of 2-Chloroacetophenones with Chiral Rhodium Catalysts. *Org. Lett.* **2002**, *4*, 4373–4376. [CrossRef] [PubMed]

23. Hamada, T.; Torii, T.; Onishi, T.; Izawa, K.; Ikariya, T. Asymmetric Transfer Hydrogenation of α-Aminoalkyl α′-Chloromethyl Ketones with Chiral Rh Complexes. *J. Org. Chem.* **2004**, *69*, 7391–7394. [CrossRef]

24. Matharu, D.S.; Morris, D.J.; Kawamoto, A.M.; Clarkson, G.J.; Wills, M. A Stereochemically Well-Defined Rhodium(III) Catalyst for Asymmetric Transfer Hydrogenation of Ketones. *Org. Lett.* **2005**, *7*, 5489–5491. [CrossRef]

25. Bevinakatti, S.; Banerji, A.A. Practical chemoenzymic synthesis of both enantiomers of propranolol. *J. Org. Chem.* **1991**, *56*, 5372–5375. [CrossRef]

26. Ader, U.; Schneider, M.P. Enzyme assisted preparation of enantiomerically pure β-adrenergic blockers III. Optically active chlorohydrin derivatives and their conversion. *Tetrahedron Asymmetry* **1992**, *3*, 521–524. [CrossRef]

27. Qin, F.; Qin, B.; Mori, T.; Wang, Y.; Meng, L.; Zhang, X.; Jia, X.; Abe, I.; You, S. Engineering of *Candida glabrata* Ketoreductase 1 for Asymmetric Reduction of α-Halo Ketones. *ACS Catal.* **2016**, *6*, 6135–6140. [CrossRef]

28. De Miranda, A.S.; Simon, R.C.; Grischek, B.; de Paula, G.C.; Horta, B.A.C.; de Miranda, L.S.M.; Kroutil, W.; Kappe, C.O.; de Souza, R.O.M.A. Chiral Chlorohydrins from the Biocatalyzed Reduction of Chloroketones: Chiral Building Blocks for Antiretroviral Drugs. *ChemCatChem* **2015**, *7*, 984–992. [CrossRef]

29. Wu, K.; Chen, L.; Fan, H.; Zhao, Z.; Wang, H.; Wei, D. Synthesis of enantiopure epoxide by 'one pot' chemoenzymatic approach using a highly enantioselective dehydrogenase. *Tetrahedron Lett.* **2016**, *57*, 899–904. [CrossRef]

30. Guo, C.; Chen, Y.; Zheng, Y.; Zhang, W.; Tao, Y.; Feng, J.; Tang, L. Exploring the Enantioselective Mechanism of Halohydrin Dehalogenase from Agrobacterium radiobacter AD1 by Iterative Saturation Mutagenesis. *Appl. Environ. Microbiol.* **2015**, *81*, 2919–2926. [CrossRef] [PubMed]

31. Tang, L.; Zhu, X.; Zheng, H.; Jiang, R.; Majerić Elenkovb, M. Key Residues for Controlling Enantioselectivity of Halohydrin Dehalogenase from Arthrobacter sp. Strain AD2, Revealed by Structure-Guided Directed Evolution. *Appl. Environ. Microbiol.* **2012**, *78*, 2631–2637. [CrossRef] [PubMed]

32. Perna, F.M.; Ricci, M.A.; Scilimati, A.; Mena, M.C.; Pisano, I.; Palmieri, L.; Agrimi, G.; Vitale, P. Cheap and environmentally sustainable stereoselective arylketones reduction by *Lactobacillus reuteri* whole cells. *J. Mol. Catal. B Enzym.* **2016**, *124*, 29–37. [CrossRef]

33. Vitale, P.; D'Introno, C.; Perna, F.M.; Perrone, M.G.; Scilimati, A. *Kluyveromyces marxianus* CBS 6556 growing cells as a new biocatalyst in the asymmetric reduction of substituted acetophenones. *Tetrahedron Asymmetry* **2013**, *24*, 389–394.

34. Vitale, P.; Perna, F.M.; Perrone, M.G.; Scilimati, A. Screening On The Use Of *Kluyveromyces Marxianus* CBS 6556 Growing Cells As Enantioselective Biocatalyst For Ketones Reduction. *Tetrahedron Asymmetry* **2011**, *22*, 1985–1993. [CrossRef]

35. Vitale, P.; Abbinante, V.M.; Perna, F.M.; Salomone, A.; Cardellicchio, C.; Capriati, V. Unveiling the Hidden Performance of Whole Cells in the Asymmetric Bioreduction of Aryl-containing Ketones in Aqueous Deep Eutectic Solvents. *Adv. Synth. Catal.* **2016**. [CrossRef]

36. Brenna, E. *Synthetic Methods for Biologically Active Molecules: Exploring the Potential of Bioreductions*; Brenna, E., Ed.; Wiley-VCH: Weinheim, Germany, 2013.

37. Faber, K. *Biotransformations in Organic Chemistry: A Textbook*, 6th ed.; Springer-Verlag: Berlin/Heidelberg, Germany, 2011.

38. Csuk, R.; Glanzer, I. *Yeast-Mediated Stereoselective Biocatalysis*; Patel, R.N., Ed.; Stereoselective Biocatalysis, Marcel Dekker: New York, NY, USA, 2000; pp. 527–578.

39. Rodrigues, J.A.R.; Moran, P.J.S.; Conceicao, G.J.A.; Fardelone, L.C. Asymmetric Reduction of Carbonyl Compounds. *Food Technol. Biotechnol.* **2004**, *42*, 295–303.

40. Utsukihara, T.; Okada, S.; Kato, N.; Horiuchi, C.A. Biotransformation of α-bromo and α,α'-dibromo alkanone to α-hydroxyketone and α-diketone by *Spirulina platensis. J. Mol. Catal. B Enzym.* **2007**, *45*, 68–72. [CrossRef]

41. Wagner, P.J.; Lindstrom, M.J.; Sedon, J.H.; Ward, D.R. Photochemistry of delta-haloketones: Anchimeric assistance in triplet-state gamma-hydrogen abstraction and beta-elimination of halogen atoms from the resulting diradicals. *J. Am. Chem. Soc.* **1981**, *103*, 3842–3849. [CrossRef]

42. Aleixo, L.M.; De Carvalho, M.; Moran, P.J.S.; Rodrigues, J.A.R. Hydride transfer versus electron transfer in the baker's yeast reduction of α-haloacetophenones. *Bioorg. Med. Chem. Lett.* **1993**, *3*, 1637–1642. [CrossRef]

43. Rocha, L.C.; Ferreira, H.V.; Pimenta, E.F.; Souza Berlinck, R.G.; Oliveira Rezende, M.O.; Landgraf, M.D.; Regali Seleghim, M.H.; Durães Sette, L.; Meleiro Porto, A.L. Biotransformation of α-bromoacetophenones by the marine fungus *Aspergillus sydowii. Mar. Biotechnol.* **2010**, *12*, 552–557. [CrossRef] [PubMed]

44. Barbasiewicz, M.; Brud, A.; Mąkosza, M. Synthesis of Substituted Tetrahydropyrans via Intermolecular Reactions of δ-Halocarbanions with Aldehydes. *Synthesis* **2007**, 1209–1213. [CrossRef]

45. De Carvalho, M.; Okamoto, M.T.; Moran, P.J.S.; Rodrigues, J.A.R. Baker's yeast reduction of α-haloacetophenones. *Tetrahedron* **1991**, *47*, 2073–2080. [CrossRef]

46. Wu, K.; Wang, H.; Chen, L.; Fan, H.; Zhao, Z.; Wei, D. Practical two-step synthesis of enantiopure styrene oxide through an optimized chemoenzymatic approach. *Appl. Microbiol. Biotechnol.* **2016**, *100*, 8757–8767. [CrossRef] [PubMed]

47. Patel, J.M.; Musa, M.M.; Rodriguez, L.; Sutton, D.A.; Popik, V.V.; Phillips, R.S. Mutation of Thermoanaerobacter ethanolicus secondary alcohol dehydrogenase at Trp-110 affects stereoselectivity of aromatic ketone reduction. *Org. Biomol. Chem.* **2014**, *12*, 5905–5910. [CrossRef] [PubMed]

48. Ricci, M.A.; Russo, A.; Pisano, I.; Palmieri, L.; de Angelis, M.; Agrimi, G. Improved 1,3-Propanediol Synthesis from Glycerol by the Robust *Lactobacillus reuteri* Strain DSM 20016. *J. Microbiol. Biotechnol.* **2015**, *6*, 893–902. [CrossRef] [PubMed]

49. De Man, J.C.; Rogosa, M.; Sharpe, M.E. A medium for the cultivation of lactobacilli. *J. Appl. Bacteriol.* **1960**, *23*, 130–135. [CrossRef]

50. Zhou, J.-N.; Fang, Q.; Hu, Y.-H.; Yang, L.-Y.; Wu, F.-F.; Xie, L.-J.; Wu, J.; Li, S. Copper(II)-catalyzed enantioselective hydrosilylation of halo-substituted alkyl aryl and heteroaryl ketones: Asymmetric synthesis of (*R*)-fluoxetine and (*S*)-duloxetine. *Org. Biomol. Chem.* **2014**, *12*, 1009–1017. [CrossRef]

51. Pop, L.A.; Czompa, A.; Paizs, C.; Toşa, M.I.; Vass, E.; Mátyus, P.; Irimie, F.-D. Lipase-Catalyzed Synthesis of Both Enantiomers of 3-Chloro-1-arylpropan-1-ols. *Synthesis* **2011**, *18*, 2921–2928.

52. Janeczko, T.; Kostrzewa-Susłow, E. Enantioselective reduction of propiophenone formed from 3-chloropropiophenone and stereoinversion of the resulting alcohols in selected yeast cultures. *Tetrahedron Asymmetry* **2014**, *25*, 1264–1269. [CrossRef]

53. Yu, F.; Zhou, J.-N.; Zhang, X.-C.; Sui, Y.-Z.; Wu, F.-F.; Xie, L.-J.; Chan, A.S.C.; Wu, J. Copper(II)-Catalyzed Hydrosilylation of Ketones Using Chiral Dipyridylphosphane Ligands: Highly Enantioselective Synthesis of Valuable Alcohols. *Chem. Eur. J.* **2011**, *17*, 14234–14240. [CrossRef] [PubMed]

54. Li, M.; Li, B.; Xia, H.-F.; Ye, D.; Wu, J.; Shi, Y. Mesoporous silica KIT-6 supported superparamagnetic CuFe₂O₄ nanoparticles for catalytic asymmetric hydrosilylation of ketones in air. *Green Chem.* **2014**, *16*, 2680–2688. [CrossRef]

55. Coppi, D.I.; Salomone, A.; Perna, F.M.; Capriati, V. 2-Lithiated-2-phenyloxetane: A new attractive synthon for the preparation of oxetane derivatives. *Chem. Commun.* **2011**, *47*, 9918–9920. [CrossRef] [PubMed]

56. Schaal, C. 2-Aryloxoetanes. I. Synthesis and NMR study. *Bull. Soc. Chim. Fr.* **1969**, *10*, 3648–3652.

57. Schaal, C. Synthese. Extension des relations lineaires d'enthalpie libre aux parametres de resonance magnetique nucleaire. *Bull. Soc. Chim. Fr.* **1971**, *8*, 3064–3070.
58. Chan, T.H.; Pellon, P. Chiral organosilicon compounds in synthesis. Highly enantioselective synthesis of arylcarbinols. *J. Am. Chem. Soc.* **1989**, *111*, 8737–8738. [CrossRef]
59. Jokisaari, J.; Rahkamaa, E.; Malo, H. Studies on the PMR Spectra of Oxetanes: IV. 2-(4-Halophenyl) oxetanes. *Zeitschrift für Naturforschung A* **1971**, *26*, 973–978. [CrossRef]
60. Zeng, X.H.; Miao, C.X.; Wang, S.F.; Xia, C.G.; Sun, W. Asymmetric 5-endo chloroetherification of homoallylic alcohols toward the synthesis of chiral β-chlorotetrahydrofurans. *Chem. Commun.* **2013**, *49*, 2418–2420. [CrossRef] [PubMed]
61. Capriati, V.; Florio, S.; Luisi, R.; Salomone, A. Oxiranyl Anion-Mediated Synthesis of Highly Enantiomerically Enriched Styrene Oxide Derivatives. *Org. Lett.* **2002**, *4*, 2445–2448. [CrossRef] [PubMed]
62. Jia, X.; Wang, Z.; Li, Z. Preparation of (*S*)-2-, 3-, and 4-chlorostyrene oxides with the epoxide hydrolase from *Sphingomonas sp.* HXN-200. *Tetrahedron Asymmetry* **2008**, *19*, 407–415. [CrossRef]

# catalysts

*Article*

# New Tailor-Made Alkyl-Aldehyde Bifunctional Supports for Lipase Immobilization

**Robson Carlos Alnoch [1,2], Ricardo Rodrigues de Melo [1,3], Jose M. Palomo [1], Emanuel Maltempi de Souza [2], Nadia Krieger [4] and Cesar Mateo [1,\*]**

[1] Departamento de Biocatálisis, Instituto de Catálisis y Petroleoquímica (CSIC), Marie Curie 2. Cantoblanco, Campus UAM, 28049 Madrid, Spain; robsonalnoch@hotmail.com (R.C.A.); ricardorodriguesmelo@gmail.com (R.R.d.M.); josempalomo@icp.csic.es (J.M.P.)
[2] Departamento de Bioquímica e Biologia Molecular, Universidade Federal do Paraná, Cx. P. 19081 Centro Politécnico, 81531-980 Curitiba, Paraná, Brazil; souzaem@ufpr.br
[3] Departamento de Ciência de Alimentos, Faculdade de Engenharia de Alimentos (FEA), Universidade Estadual de Campinas, 13083-862 Campinas, São Paulo, Brazil
[4] Departamento de Química, Universidade Federal do Paraná, Cx. P. 19081 Centro Politécnico, 81531-980 Curitiba, Paraná, Brazil; nkrieger@ufpr.br
\* Correspondence: ce.mateo@icp.csic.es; Tel.: +34-915854768; Fax: +34-915854860

Academic Editor: David D. Boehr
Received: 28 October 2016; Accepted: 27 November 2016; Published: 30 November 2016

**Abstract:** Immobilized and stabilized lipases are important biocatalytic tools. In this paper, different tailor-made bifunctional supports were prepared for the immobilization of a new metagenomic lipase (LipC12). The new supports contained hydrophobic groups (different alkyl groups) to promote interfacial adsorption of the lipase and aldehyde groups to react covalently with the amino groups of side chains of the adsorbed lipase. The best catalyst was 3.5-fold more active and 5000-fold more stable than the soluble enzyme. It was successfully used in the regioselective deacetylation of peracetylated D-glucal. The PEGylated immobilized lipase showed high regioselectivity, producing high yields of the C-3 monodeacetylated product at pH 5.0 and 4 °C.

**Keywords:** regioselective hydrolysis; biocatalysis; lipase; interfacial activation; covalent immobilization; tailor-made supports; enzyme stabilization

## 1. Introduction

Lipases (EC 3.1.1.3) normally catalyze the hydrolysis of carboxylic esters in aqueous media, but they can also be used to synthesize carboxylic esters in water-restricted media, exhibiting high regio-, chemo-, and enantioselectivity. Due to these properties, lipases have been used in different reactions, standing out among the most widely used enzymes in biotechnology [1,2].

Recently, a new lipase, LipC12, was identified in a metagenomic library constructed from soil samples contaminated with fat [3]. LipC12 had a specific activity against long-chain triglycerides (e.g., olive oil 1722 U·mg$^{-1}$) that is comparable to the specific activities of several well-known commercial lipases [3,4]. Furthermore, LipC12 was stable at moderate temperatures and in the presence of co-solvents such as methanol, propanol, or acetone [3]. These features suggest that LipC12 might be suitable for use in biocatalysis.

Typically, industrial biocatalytic processes require that the lipases be immobilized since that immobilization facilitates reutilization of the enzyme, reducing process costs [5]. The immobilization of different lipases has been performed using different immobilization methods such as covalent linkage with different reactive groups, electrostatic or hydrophobic adsorptions, entrapment, encapsulation, or cross-linked enzyme aggregates (CLEAs); and using different materials such as nanomagnetic particles, microspheres, organic or inorganic materials, porous and/or macroporous gel beads, graphene oxides, exfoliated bentonite, and many others [6–19]. The different immobilization methods have enabled the procurement of lipase catalysts with different properties in terms of activity, stability, and selectivity [5].

The most successful strategy used for lipase immobilization is adsorption on hydrophobic supports [5,20]. This strategy has permitted the purification and immobilization of various lipases in a single step [9,11,21,22]. These protocols are based on the special characteristics and mechanisms of lipases. In aqueous media, lipases are in equilibrium between closed and open forms. In the closed form, the lid, which is formed by a short alpha helix, secludes the catalytic site from the medium, making it inaccessible to the substrate, such that the lipase is in an inactive state. In the open and active form, the internal side of the lid and the surroundings of the active site form a hydrophobic pocket that is exposed to the medium. The open form is stabilized upon contact of the lipase with a hydrophobic surface, as occurs at the oil-water interface when lipases are used to hydrolyze triacylglycerides in oil-in-water emulsions [9,11]. The adsorption of lipases in the open form at this interface leads to high activity in a phenomenon that is called interfacial activation. Immobilization by adsorption on hydrophobic surfaces takes advantage of this phenomenon by fixing the lipase predominantly in its open conformation. This gives this method a significant advantage over other methods that immobilize the lipase by other regions and which therefore allow the immobilized lipase to equilibrate between the open and closed conformations. This method is specific and yields more active and selective catalysts [23,24], this being especially important for the catalysis of complex reactions, such as regioselective deprotection reactions with carbohydrates [25]. However, physical adsorption also has a significant disadvantage: the association between the protein and the support is reversible, meaning that the lipase can leach from the solid support, especially in the presence of low concentrations of detergents or solvents [9].

One strategy for preventing the leaching of lipases from hydrophobic supports would be to create covalent bonds between the adsorbed enzyme and the support. In fact, covalent immobilization of enzymes using aldehyde-activated supports is a widely used technique [26]. However, a heterofunctional support that combines hydrophobic and aldehyde groups in the same matrix has not previously been described.

In the present work, novel tailor-made alkyl-aldehyde supports were prepared (Scheme 1). The novel supports contain: (i) a very dense layer of different hydrophobic moieties (different alkyl groups) that are able to absorb lipases at neutral pH; and (ii) a high concentration of aldehyde groups that are able to react covalently with the enzyme, especially at alkaline pH. The presence of different groups with different functions on the surface of the support should permit better control of the immobilization, which occurs through a two-step mechanism: first the enzyme adsorbs onto the hydrophobic groups and then the aldehyde groups react with it, immobilizing it covalently.

These novel functionalized supports were used to immobilize the novel lipase LipC12, and the stability, activity, and regioselectivity of the new heterogeneous biocatalyst were tested. The best heterogeneous biocatalyst that was obtained was used in the regioselective hydrolysis of per-*O*-acetylated D-glucal, an interesting building block for the synthesis of various tailor-made di- and trisaccharides.

A

B

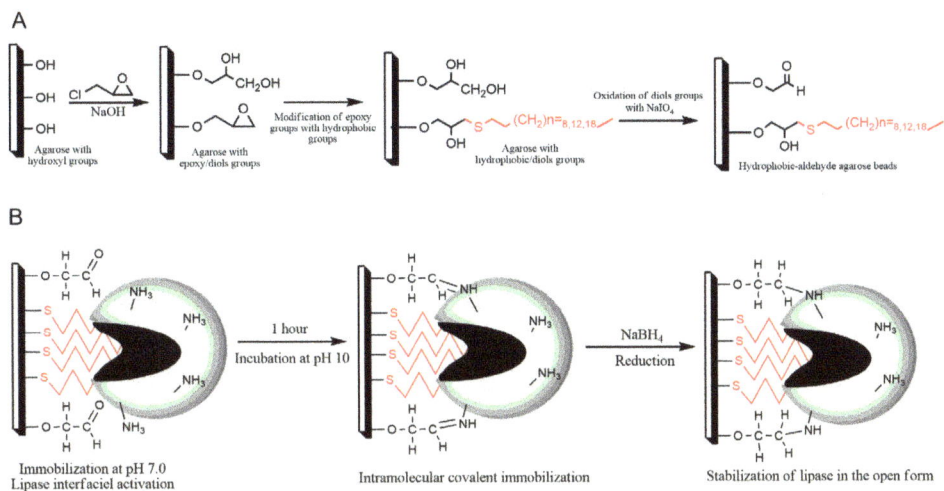

**Scheme 1.** (**A**) Preparation of new tailor-made alkyl-aldehyde supports; (**B**) Mechanism of immobilization-stabilization of lipases in the open form on new alkyl-aldehyde supports. $n$ = C8 (1-octanethiol); C12 (1-dodecanethiol) and C18 (1-octadecanethiol).

## 2. Results and Discussion

### 2.1. Preparation of New Alkyl-Aldehyde Supports

Agarose beads were utilized as the base matrix for the construction of different bifunctional supports. The surface of the support, which is rich in primary hydroxyl groups, was activated in alkaline conditions, with epiclorohydrin, forming epoxy groups and diol groups (Scheme 1A). The total amount of activated primary hydroxyl groups was around 65 $\mu$mol·g$^{-1}$, with epoxy groups accounting for 23 $\mu$mol·g$^{-1}$ and diol groups accounting for 42 $\mu$mol·g$^{-1}$ (Table 1). The epoxy groups were functionalized with different bifunctional hydrophobic agents (octane-, dodecane-, and octadecane-thiol) in order to have supports containing groups with different degrees of hydrophobicity for interfacial adsorption of the lipase (Scheme 1A). The diol groups were then oxidized with sodium periodate, producing aldehyde groups. These aldehyde groups are capable of reacting covalently with different amine groups of the protein. Immobilization of the lipase on this support occurs in two steps: first, the enzyme adsorbs hydrophobically in an orientation that favors the open form; second, the aldehyde groups react covalently with the side chains of lysine that are exposed at the surface of the enzyme, fixing it covalently o the support (Scheme 1B).

**Table 1.** Quantification of groups on the new alkyl-aldehyde supports.

| Support | Ligands ($\mu$mol·g$^{-1}$) | Diol Groups ($\mu$mol·g$^{-1}$) |
|---|---|---|
| Agarose-Epoxy | 23 $\pm$ 0.4 | 43 $\pm$ 0.4 |
| C8-aldehyde | 23 $\pm$ 1 | 43 $\pm$ 1 |
| C12-aldehyde | 21 $\pm$ 1.6 | 41 $\pm$ 1.6 |
| C18-aldehyde | 19 $\pm$ 1.1 | 38 $\pm$ 1.1 |

The number of epoxy/ligands groups was calculated from the difference in periodate consumption between the hydrolyzed support and the initial epoxy support as described in the methods section. Results are expressed as the average of triplicate assays $\pm$ the standard error of the mean.

## 2.2. Immobilization of LipC12 on New Alkyl-Aldehyde Supports

Figure 1 shows the immobilization of LipC12 by adsorption onto the new alkyl-aldehyde supports. LipC12 was quite rapidly immobilized at pH 7.0 on all bifunctionalized supports, with complete immobilization (i.e., >95% removal of activity from the supernatant) occurring in less than 2 h.

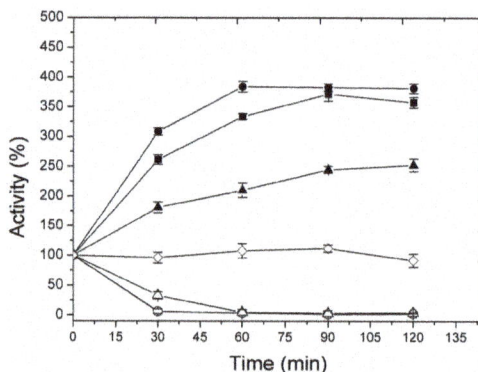

**Figure 1.** Immobilization courses of LipC12 on new alkyl-aldehyde supports. (□) C8-aldehyde; (○) C12-aldehyde; (△) C18-aldehyde. (◇) Control. Symbols: Black (suspension); Hollow (supernatant). Results are expressed as the average of triplicate assays ± the standard error of the mean.

LipC12 was activated by adsorption onto the support, with the activities measured for the suspension being significantly higher than that of the original supernatant (Figure 1). The highest value of recovered activity, 380%, was obtained with the preparation C12-aldehyde/LipC12. The preparations C8-aldehyde/LipC12 and C18-aldehyde/LipC12 also showed high values of recovered activity (>200%), showing the hyperactivation of lipase LipC12 immobilized these supports (Table 2). The results show that these new tailor-made supports allowed the immobilization of this lipase in its open conformation via interfacial activation [9,11].

**Table 2.** Principal parameters for immobilization of the lipase LipC12 on new alkyl-aldehyde supports.

| Support | Immobilization Efficiency (%) [a] | Recovered Activity [b] (%) | Recovered Activity after Reduction [c] |
|---------|-----------------------------------|----------------------------|----------------------------------------|
| C8-aldehyde | >95 | 357 | 346 |
| C12-aldehyde | >95 | 380 | 370 |
| C18-aldehyde | >95 | 252 | 256 |

[a] Calculated as the difference between the initial and final activities in the supernatant after 2 h of immobilization; [b] Recovered activity (%), measured as the ratio between the real activity ($U \cdot g^{-1}$ support) of immobilized LipC12 and theoretical activity of the immobilized LipC12 ($U \cdot g^{-1}$ support); [c] Recovered activity (%) after incubation at pH 10 for 1 h and reduction with $NaBH_4$.

In order to fix LipC12 covalently to the support, the immobilized preparations were incubated at different pH values (7.0, 8.5, and 10) for 1 h. After the incubation, the imine bonds formed between the enzyme and the support were then reduced by adding sodium borohydride. This reduction did not affect the activity of the immobilized enzyme (Table 2). No leaching of lipase was found after incubation in surfactants.

## 2.3. Thermal Inactivation of Different Immobilized LipC12 Preparations

The various immobilized LipC12 preparations previously incubated at different pH values were incubated in phosphate buffer 25 mM at 55 °C. In all cases, the thermal stability of the derivatives incubated at pH 10 was higher than that incubated at pH 8.5 and 7.0 or the only adsorbed preparations

(Figure S1). At pH 7.0, the reactivity of the amino groups of the enzymes was not high enough to produce a covalent attachment with the aldehyde groups; at pH 10, the increase in the reactivity of the amine groups of side chains close to the lid that promote the rigidification on this region resulting in a high stabilization. At 55 °C, C8-aldehyde/LipC12, C12-aldehyde/LipC12, C18-aldehyde/LipC12 conserved more than 80% of their activity after 24 h (Figure 2A) while the half-life of the soluble enzyme was 37 min.

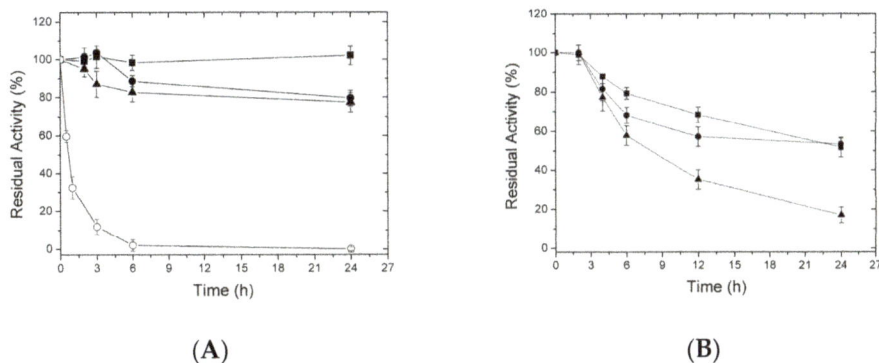

(A)                                        (B)

**Figure 2.** Thermal inactivation of LipC12 immobilized on different alkyl-aldehyde supports. (**A**) Inactivation was performed at pH 7.0, 55 °C after incubation at pH 10 for 1 h; (**B**) Inactivation was performed at pH 7.0, 80 °C after incubation at pH 10 for 1 h. (■) C12-aldehyde/LipC12; (●) C8-aldehyde/LipC12; (▲) C18-aldehyde/LipC12 and (○) Soluble enzyme. Results are expressed as the average of triplicate assays ± the standard error of the mean.

After 24 h incubation at 80 °C, C8-aldehyde/LipC12 and C12-aldehyde/LipC12 still had residual activities above 50%, while the residual activity of C18-aldehyde/LipC12 was only 20% (Figure 2B). Intermediary spacer arms (C8 and C12) supports produced a slight increment in the stability effect achieved when compared with C18. The half-lives were 22 h for C8-aldehyde/LipC12 and 21 h for C12-aldehyde/LipC12, while the soluble lipase lost 50% of the activity after only 15 s. This means that the alkyl-aldehyde-lipase preparations were from 2000- to 5000-fold more stable than the soluble enzyme (Table 3). Considering the retention of activity (Table 2) and stability, the C12-aldehyde/LipC12 preparation was chosen for the remaining studies.

**Table 3.** Half-lives (in hours) of the different immobilized preparations at 80 °C.

| Preparations [a] | Half-Life ($t\frac{1}{2}$) at 80 °C | Stability Factor |
|---|---|---|
| Soluble enzyme | 0.004 | - |
| C8-aldehyde/LipC12 | 22 | 5500 |
| C12-aldehyde/LipC12 | 21 | 5250 |
| C18-aldehyde/LipC12 | 8 | 2000 |

[a] Preparations were incubated at 80 °C. Aliquots were withdrawn periodically for quantification of residual enzymatic activity to estimate the half-life according to Henley and Sadana [27].

### 2.4. Effect of Temperature and pH on Activity of Free and Immobilized LipC12

The optimum temperatures for the activity of free and immobilized LipC12 were determined over the temperature range of 20–90 °C. The maximum activity of the free enzyme was obtained at 30 °C while the optimal temperature for C12-aldehyde/LipC12 was 70 °C (Figure 3).

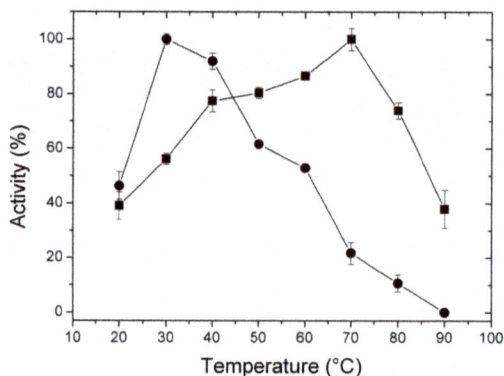

**Figure 3.** Effect of temperature on free and C12-aldehyde/LipC12 activity. (●) Soluble LipC12; (■) C12-aldehyde/LipC12. The activity was determined using p-nitrophenyl proprionate (pNPP) as the substrate, at pH 7.0. Results are expressed as the average of triplicate assays ± the standard error of the mean.

This shift in the optimal temperature was related to the improvement of the stability of the obtained preparation. The high improvement after adsorption and covalent linkage is important because it permits the transformation of a mesophilic enzyme into an enzyme with properties that are similar to, or even better than, those of enzymes from thermophile organisms, such as Bacillus thermocatenolatus lipase (BTL) and Thermus thermophilus lipase (TTL) [28,29].

In relation to the effects of pH on activity, the maximum activity was obtained at around pH 7.0 for both free LipC12 and C12-aldehyde/LipC12 (Figure 4).

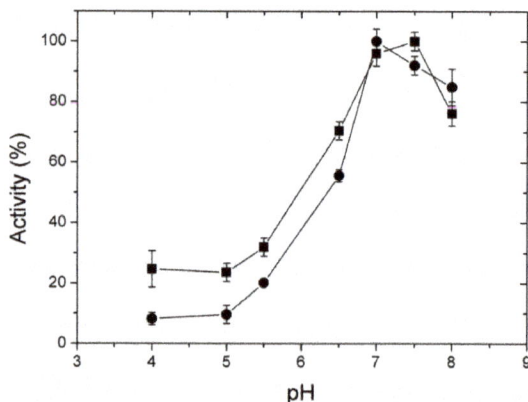

**Figure 4.** Effect of pH on free and C12-aldehyde/LipC12 activity. (●) Soluble LipC12; (■) C12-aldehyde/LipC12. The activity was determined using p-nitrophenyl proprionate (pNPP) as the substrate. Results are expressed as the average of triplicate assays ± the standard error of the mean.

### 2.5. Regioselective Hydrolysis of 3,4,6-tri-O-acetyl-D-glucal by Immobilized LipC12

C12-aldehyde/LipC12 was used to catalyze the hydrolytic deacetylation of per-O-acetylated-D-glucal (1). The yield of this reaction depends strongly on the reaction conditions. The principal variables assayed were the pH and temperature. Additionally, the recovering of the optimal catalyst with PEG was performed. This treatment has demonstrated that it is able to improve the activity and stability [30].

The activity of the soluble enzyme was also assayed. However, at 25 °C, its activity was extremely low, so no attempt was made to assay it at 4 °C (data not shown).

At 25 °C, low regioselectivity was C12-aldehyde/LipC12 at both pH 7.0 and pH 5.0, producing only around 10% yield of monodeacetylated products at 100% conversion (Table 4). The PEGylated preparation, C12-aldehyde/LipC12-PEG, had a slightly improved regioselectivity at pH 5.0 and 25 °C, although the yield of 3-OH product (**2**) was only 22%.

**Table 4.** Regioselective hydrolysis of 3,4,6-tri-*O*-acetyl-D-glucal (**1**) using C12-aldehyde/LipC12.

| Preparation | pH | T °C | Specific Activity (U·mg$^{-1}$) * | Time (h) | Total Conversion [a] (%) | Yield 2 (%) | Yield 3 (%) | Yield 4 (%) | Other Products [b] (%) |
|---|---|---|---|---|---|---|---|---|---|
| C12-aldehyde /LipC12 | 7.0 | 4 | 18 | 96 | 77 | 52 | 1 | 7 | 17 |
| C12-aldehyde /LipC12 | 5.0 | 4 | 4 | 96 | 34 | 26 | 1 | 1 | 6 |
| C12-aldehyde /LipC12-PEG | 5.0 | 4 | 15 | 96 | 81 | 69 | 0 | 3 | 9 |
| C12-aldehyde /LipC12 | 7.0 | 25 | 140 | 24 | 100 | 5 | 4 | 0 | 91 |
| C12-aldehyde /LipC12 | 5.0 | 25 | 140 | 24 | 100 | 11 | 0 | 0 | 89 |
| C12-aldehyde /LipC12-PEG | 5.0 | 25 | 110 | 24 | 100 | 22 | 2 | 2 | 74 |

(**1**)- 3,4,6-tri-*O*-acetyl-D-glucal; (**2**)- 4,6-di-*O*-acetyl-D-glucal; (**3**)- 3,4-di-*O*-acetyl-D-glucal; and (**4**)- 3,6-di-*O*-acetyl-D-glucal; * ×10$^{-3}$; [a] Total conversion of substrate (**1**) with different products; [b] D-glucal and dideacetylated products.

The regioselectivity was higher at 4 °C than at 25 °C (Table 4). At pH 7.0, 52% of C-3-OH product (**2**) was obtained at 77% conversion, with slight conversion into 4-OH product (**4**) (7%) and 6-OH product (**3**) (1%), reducing the undesired product in 17% (Table 4). The PGEylation of this catalyst (C12-aldehyde/LipC12-PEG) allowed an improvement of the regioselectivity. This catalyst produced 69% of 3-OH product (**2**) at 81% conversion, and only 3% of 4-OH product (**4**) (Table 4).

The PEGylated catalyst was reused in three reaction cycles at 4 °C and similar reaction yields were obtained, demonstrating its reusability (Figure S2). However, the recycle of the catalysts in this reaction are not reported, these data are similar to others obtained by different authors for the hydrolysis of esters as reported by Macario et al. [31], where the catalyst (lipase of *Rhizomucor miehei* immobilized on zeolites) was used in the hydrolysis of methyl myristate for four cycles or Cao et al. [12] that recycled the catalyst (nanohybrids of *Yarrawia* lipolytica lipase) for 12 reaction cycles using *p*NPP as substrate.

## 3. Materials and Methods

### 3.1. Materials

The strains *E. coli* TOP10 (Invitrogen, Carlsbad, CA, USA) and BL21(DE3) (Novagen, Madison, MI, USA) and the vector pET-28a(+) (Novagen, Madison, MI, USA) were used as the recombinant protein expression system. Agarose 4 BCL was purchased from Agarose Bead Technologies (Madrid, Spain). Epichlorhydrine, iminodiacetic acid, triethylamine, sodium borohydride, sodium periodate, 1-octanothiol, 1-dodecanothiol, 1-octadecanethiol, tri-*O*-acetyl-D-glucal, polyethylene glycol (1.500), nickel(II) chloride hexahydrate, and high molecular weight protein (Sigma Marker™) were purchased from Sigma (Sigma-Aldrich, St. Louis, MO, USA). The substrate *p*-nitrophenyl proprionate (*p*NPP) was synthesized according to Ghosh et al. [32]. All other chemicals used were of analytical grade.

### 3.2. Overexpression of Recombinant LipC12

*E. coli* BL21(DE3) cells carrying the pET28a(+)/lipC12 plasmid were grown in 500 mL of LB medium at 37 °C until an $OD_{600}$ of 0.5 and induced by the addition of Iso-propyl β-D thiogalactopyranoside (IPTG) to a final concentration of 0.5 mM. The induced culture was incubated for a further 16 h at 20 °C before harvesting the cells by centrifugation (10,000 rpm for 5 min) at 4 °C. The cell pellet was re-suspended in 30 mL of lysis buffer (50 mM Tris-HCl pH 7.5, 500 mM NaCl, 10 mM β-mercaptoethanol, 1% (*v/v*) Triton X-100 and 10% (*v/v*) glycerol) and disrupted by ultrasonication in an ice bath (15 cycles of 20-s pulses, 90 W, with 30-s intervals), using a SONICATOR® XL 2020 (Heat Systems-Ultrasonics Inc., New Highway, Farmingdale, NY, USA). The crude extract was then centrifuged at 15,000 rpm 30 min at 4 °C to pellet the cell debris.

### 3.3. Protein Content Determination and Electrophoresis Analysis

Protein content was determined by the Bradford method [33] using a Coomassie Protein Assay Kit (Pierce Biotechnology, Rockford, IL, USA) with bovine serum albumin as the standard. Electrophoresis of protein samples was done with 12% (*w/v*) SDS-PAGE [34] and the gel was stained with Coomassie Brilliant Blue R-250 and destained with methanol/acetic-acid/water (5/1/4 *v/v/v*). A mixture of high molecular weight proteins (Sigma Marker™, Sigma-Aldrich®) was used as the molecular weight standard.

### 3.4. Lipase Activity Assay

Lipase activity was determined using *p*-nitrophenyl proprionate (*p*-NPP) as the substrate. Free or immobilized enzyme was added to the reaction mixture (0.4 mM *p*NPP, mM $NaH_2PO_4$ pH 7.0) and the increase of absorbance was monitored at 348 nm (at pH 7, $\varepsilon_{348\,nm}$ = 5150 $M^{-1} \cdot cm^{-1}$) [35]. One unit of activity (U) was defined as the production of 1 μmoL of *p*-nitrophenol per minute, under the assay conditions.

### 3.5. Preparation of Supports

#### 3.5.1. Epoxy-Agarose

The epoxy-agarose support was prepared according to Mateo et al. [26]. Briefly, 10 g of agarose BLC (cross-linked 4% agarose beads) was mixed with 44 mL of distilled water, 3.2 g of NaOH, 200 mg of $NaBH_4$, 16 mL of acetone, and 11 mL of epichlorohydrin. The suspension was stirred for 16 h at 25 °C. The epoxy-agarose was washed with an excess of water, filtered through a glass filter, and stored at 4 °C.

#### 3.5.2. Epoxy-Agarose-IDA-$Ni^{2+}$

The epoxy-agarose support was treated with 0.5 M iminodiacetic acid in solution at pH 11 for different durations (1, 3, 5, and 24 h), 25 °C. The support was then chelated with a $NiCl_2$ solution (30 mg·$mL^{-1}$) for 1 h. Finally, the support was washed, filtered under using a glass filter, and stored at 4 °C.

#### 3.5.3. Alkyl-Agarose-Aldehyde

The epoxy-agarose support was treated with 100 mM of different alkyl thiols (1-octanothiol; 1-dodecanothiol and 1-octadecanethiol) in a 25 mM $NaHCO_3$ solution at pH 10 for 24 h, 25 °C. For the treatment with 1-octadecanethiol, 50% (*v/v*) acetone was used as a co-solvent. The reagent was solubilized using a 50:50 (*v/v*) mixture of acetone and $NaHCO_3$ solution. After that, the supports were oxidized with $NaIO_4$ (100 mM), washed, filtered through a glass filter, and stored at 4 °C.

The number of epoxy/ligand groups was calculated from the difference in periodate consumption between the hydrolyzed support and the initial epoxy support. Periodate consumption was quantified using potassium iodide, as previously described [36].

### 3.6. Purification of Recombinant LipC12

The purification was performed using the IDA-Ni$^{2+}$ supports prepared from agarose gel beads and activated with different amounts of metal chelate groups [37]. The optimal support was that obtained after 3 h of activation with IDA (data not shown). For the purification, 4 mL of crude extract (3.2 mg·mL$^{-1}$) was offered for 1 g of support and the residual activity of the supernatant was monitored over time. After that, the support was washed three times with 25 mM NaH$_2$PO$_4$ pH 7.0 and resuspended in the same buffer at increasing concentrations of imidazole. Figure S3 shows the protein band corresponding to the molecular mass of LipC12 (32 kDa) after SDS-PAGE of the eluate from IDA-Ni$^{2+}$ support at 50 mM of imidazole. Table S1 summarizes the results of the purification step, showing an activity yield of 58%. The specific hydrolytic activity against *p*NPP was 6.2 U·mg$^{-1}$. This preparation was used in further experiments of immobilization.

### 3.7. Enzyme Immobilization

A standard protocol was established for the immobilization of LipC12 on all supports. One gram of support was suspended in 4 mL of enzyme solution (containing 0.6 mg of protein) in 25 mM NaH$_2$PO$_4$ at pH 7, 25 °C and left under mild stirring. The time course of immobilization was evaluated by determining the activity (Section 3.4) in aliquots of the supernatant and suspension removed over time. After the immobilization, the preparations were washed with 25 mM NaH$_2$PO$_4$ pH 7.0 and incubated in 4 mL of 25 mM NaHCO$_3$ at different pH values (7.0, 8.5, 10) at 25 °C for 1 h. Finally, the preparations were reduced by adding NaBH$_4$ (1 mg·mL$^{-1}$) at pH 10 and leaving the mixture under stirring for 30 min.

The immobilization efficiency (IE, %) was calculated as:

$$EI = \frac{A_i - A_f}{A_i} \times 100\% \tag{1}$$

where $A_i$ is the hydrolytic activity (U) of the enzyme solution before immobilization and $A_f$ is the hydrolytic activity (U) remaining in the supernatant at the end of the immobilization procedure.

The recovered activity (R, %) was calculated as:

$$R = \frac{A_o}{A_T} \times 100\% \tag{2}$$

where $A_0$ is the observed hydrolytic activity the immobilized preparation (U·g$^{-1}$ of support) and $A_T$ is the theoretical activity of the immobilized preparation (U·g$^{-1}$ of support), calculated based on the amount of activity removed from the supernatant during the immobilization procedure.

In some assays, immobilized preparations were treated after reduction with PEG (polyethylene glycol). PEG was used as an additive due to its protective effect on the enzymes described in the literature [30,38]. To assay, 1 g of immobilized preparation was added to 10 mL phosphate buffer pH 7.0 25 mM containing 40% PEG1500 (*w/v*). The suspension was stirred for 2 h at 25 °C. After that, the preparation was washed, filtered under using a glass filter, and stored at 4 °C.

### 3.8. Thermal Stability

The thermal stabilities of free and immobilized LipC12 were assessed by incubation in sodium phosphate buffer (25 mM, pH 7.0) in a water bath at 55 and 80 °C. Inactivation was modeled based on the deactivation theory proposed by Henley and Sadana [19]. Inactivation parameters were determined from the best-fit model of the experimental data which was the one based on a two-stage series inactivation mechanism with residual activity. Half-life was used to compare the stability of the different preparations, being determined by interpolation from the respective models described in [39].

### 3.9. Effect of pH and Temperature on the Activity of Free and Immobilized LipC12

The optimum temperature for the activity of free and immobilized LipC12 was determined over the temperature range of 20–90 °C. The effect of pH on the activity was determined over a range of pH 4.0–8.0, at 25 °C, using citrate (pH 4.0–6.0) and phosphate (pH 6.0–8.0) buffers at 25 mM. The activity was determined using *p*-nitrophenyl proprionate (*p*NPP) as substrate (Section 3.4). The activities were calculated in relation to controls that were treated identically, but without enzyme to control of spontaneous hydrolysis of the substrate.

### 3.10. Hydrolysis of 3,4,6-tri-O-acetyl-D-glucal

For the hydrolysis of peracetylated 3,4,6-tri-*O*-acetyl-D-glucal, 200 mg of immobilized Lipc12 was added to a solution (1.5 mL) of substrate-1 (1 mM) in 25 mM of phosphate (pH 7.0) or acetate (pH 5.0) buffer. The reaction was carried out at 25 or 4°C, 50 rpm. Samples were removed and analyzed by reverse phase HPLC (Spectra Physic SP 100, Thermo Fisher-Scientific, Waltham, MA, USA) using a Kromasil C18 column (25 cm × 0.4 cm, 5 μm·Ø) and a UV detector (Spectra Physic SP 8450, Thermo Fisher-Scientific, Waltham, MA, USA) set at 220 nm. The mobile phase utilized was acetonitrile (20%) in milli-Q water. The products were characterized and identified as previously described in [24]. Retention times were: 3,4,6-tri-*O*-acetyl-D-glucal **1**-24.6 min, C-3 monodeacetylated **2**-6.3 min, C-6 monodeacetylated **3**-6.6 min and C-4 monodeacetylated **4**-8.1 min. One unit of activity (U) was defined as the hydrolysis of 1 μmol of substrate per hour. Activities were expressed as specific activities (U per mg of immobilized protein). The reutilization of immobilized preparations was studied using the same reaction conditions as described above.

## 4. Conclusions

Bifunctional supports with aldehyde and different hydrophobic groups have been synthesized. The main advantage of the immobilization protocol developed in the current work is the ease with which the amounts of aldehyde and hydrophobic groups on the surface of the support can be controlled. This enables modulation of immobilization conditions which may be adapted to the immobilization/stabilization of proteins which may be limited in commercial supports. This versatile strategy could also be applied to synthesize supports with other hydrophobic groups to immobilize different lipases, producing catalysts with different properties. These modulated lipase biocatalysts could be used to produce products that are difficult synthesize by traditional methods.

The use of different supports allowed us to obtain immobilized preparations of LipC12 with different activities and stabilities. The best catalyst was 3.5-fold more active and 5000-fold more stable than the soluble enzyme. Thus, the immobilization procedure converted a mesophilic enzyme into an enzyme that can operate at high temperature, with a maximal activity obtained at 70 °C.

The optimal catalyst was used for the regioselective hydrolysis of peracetylated-D-Glucal. The highest yield of the C-3 monodeacetylated product was 69% with a conversion of 81%, at pH 5 and 4 °C using the PEGylated preparation.

**Supplementary Materials:** The following are available online at www.mdpi.com/2073-4344/6/12/191/s1, Figure S1: SDS-PAGE analyses of the LipC12 purification; Figure S2: Thermal stability of different preparations of LipC12; Figure S3: Hydrolysis of 3,4,6-tri-*O*-acetyl-D-glucal during successive reaction cycles; Table S1: Summary of the purification of LipC12.

**Acknowledgments:** The authors gratefully acknowledge the Ramón Areces Foundation for financial support. Research scholarships were granted to Robson Carlos Alnoch and Ricardo Rodrigues de Melo (Grant No: 201757/2015-0 and 201688/2015-8) for the development of personnel in higher education, and to Nadia Krieger and Emanuel Maltempi de Souza by CNPq (Conselho Nacional de Desenvolvimento Cientifico e Tecnológico), a Brazilian government agency for the advancement of science. The authors thank David A. Mitchell for critical review of the manuscript.

**Author Contributions:** R.C.A. and R.R.d.M performed the experiments and partially wrote the paper. J.M.P. contributed in the design of the reaction and partially wrote the paper. E.M.S. and N.K. prepared the enzyme and partially wrote the paper. C.M. conceived and designed the experiments and partially wrote the paper.

**Conflicts of Interest:** The authors declare no conflict of interest.

## References

1. Jaeger, K.-E.; Ransac, S.; Dijkstra, B.W.; Colson, C.; van Heuvel, M.; Misset, O. Bacterial lipases. *FEMS Microbiol. Rev.* **1994**, *15*, 29–63. [CrossRef] [PubMed]
2. Kapoor, M.; Gupta, M.N. Lipase promiscuity and its biochemical applications. *Process Biochem.* **2012**, *47*, 555–569. [CrossRef]
3. Glogauer, A.; Martini, V.P.; Faoro, H.; Couto, G.H.; Müller-Santos, M.; Monteiro, R.A.; Mitchell, D.A.; de Souza, E.M.; Pedrosa, F.O.; Krieger, N. Identification and characterization of a new true lipase isolated through metagenomic approach. *Microb. Cell Fact.* **2011**, *10*. [CrossRef] [PubMed]
4. Hasan, F.; Shah, A.A.; Hameed, A. Industrial applications of microbial lipases. *Enzyme Microb. Technol.* **2006**, *39*, 235–251. [CrossRef]
5. Santos, J.C.S.D.; Barbosa, O.; Ortiz, C.; Berenguer-Murcia, A.; Rodrigues, R.C.; Fernandez-Lafuente, R. Importance of the Support Properties for Immobilization or Purification of Enzymes. *ChemCatChem* **2015**, *7*, 2413–2432. [CrossRef]
6. Alnoch, R.C.; Martini, V.P.; Glogauer, A.; Costa, A.C.D.S.; Piovan, L.; Muller-Santos, M.; De Souza, E.M.; Pedrosa, F.D.O.; Mitchell, D.A.; Krieger, N. Immobilization and characterization of a new regioselective and enantioselective lipase obtained from a metagenomic library. *PLoS ONE* **2015**, *10*, e0114945. [CrossRef] [PubMed]
7. Rueda, N.; Dos Santos, C.S.; Rodriguez, M.D.; Albuquerque, T.L.; Barbosa, O.; Torres, R.; Ortiz, C.; Fernandez-Lafuente, R. Reversible immobilization of lipases on octyl-glutamic agarose beads: A mixed adsorption that reinforces enzyme immobilization. *J. Mol. Catal. B Enzym.* **2016**, *128*, 10–18. [CrossRef]
8. Guisan, J.M.; Sabuquillo, P.; Fernandez-Lafuente, R.; Fernandez-Lorente, G.; Mateo, C.; Halling, P.J.; Kennedy, D.; Miyata, E.; Re, D. Preparation of new lipases derivatives with high activity-stability in anhydrous media: Adsorption on hydrophobic supports plus hydrophilization with polyethylenimine. *J. Mol. Catal. B Enzym.* **2001**, *11*, 817–824. [CrossRef]
9. Palomo, J.M.; Munoz, G.; Fernandez-Lorente, G.; Mateo, C.; Fernandez-Lafuente, R.; Guisan, J.M. Interfacial adsorption of lipases on very hydrophobic support (octadecyl-Sepabeads): immobilization, hyperactivation and stabilization of the open form of lipases. *J. Mol. Catal. B Enzym.* **2002**, *19*, 279–286. [CrossRef]
10. Carrasco-Lopez, C.; Godoy, C.; de las Rivas, B.; Fernandez-Lorente, G.; Palomo, J.M.; Guisan, J.M.; Fernandez-Lafuente, R.; Martinez-Ripoll, M.; Hermoso, J.A. Activation of bacterial thermo alkalophilic lipases is spurred by dramatic structural rearrangements. *J. Biol. Chem.* **2009**, *284*, 4365–4372. [CrossRef] [PubMed]
11. Fernandez-Lafuente, R.; Armisen, P.; Sabuquillo, P.; Fernández-Lorente, G.; Guisán, J.M. Immobilization of lipases by selective adsorption on hydrophobic supports. *Chem. Phys. Lipids* **1998**, *93*, 185–197. [CrossRef]
12. Cao, J.; Li, Y.; Tu, N.; Lv, Y.; Chen, Q.; Dong, H. Novel enzyme/exfoliated bentonite nanohybrids as highly efficient and recyclable biocatalysts in hydrolytic reaction. *J. Mol. Catal. B Enzym.* **2016**, *132*, 41–46. [CrossRef]
13. Chen, C.; Zhu, X.; Gao, Q.; Fang, F.; Wang, L.; Huang, X. Immobilization of lipase onto functional cyclomatrix polyphosphazene microspheres. *J. Mol. Catal. B Enzym.* **2016**, *132*, 67–74. [CrossRef]
14. Sato, R.; Tokuyama, H. Fabrication of enzyme-entrapped composite and macroporous gel beads by suspension gelation combined with sedimentation polymerization. *Biochem. Eng. J.* **2016**, *113*, 152–157. [CrossRef]
15. Isobe, N.; Lee, D.S.; Kwon, Y.J.; Kimura, S.; Kuga, S.; Wada, M.; Kim, U.J. Immobilization of protein on cellulose hydrogel. *Cellulose* **2011**, *18*, 1251–1256. [CrossRef]
16. Verri, F.; Diaz, U.; MacArio, A.; Corma, A.; Giordano, G. Optimized hybrid nanospheres immobilizing Rhizomucor miehei lipase for chiral biotransformation. *Process Biochem.* **2016**, *51*, 240–248. [CrossRef]

17. Mathesh, M.; Luan, B.; Akanbi, T.O.; Weber, J.K.; Liu, J.; Barrow, C.J.; Zhou, R.; Yang, W. Opening Lids: Modulation of Lipase Immobilization by Graphene Oxides. *ACS Catal.* **2016**, *6*, 4760–4768. [CrossRef]

18. Yagar, H.; Balkan, U. Entrapment of laurel lipase in chitosan hydrogel beads. *Artif. Cells Nanomed. Biotechnol.* **2016**, 1–7. [CrossRef] [PubMed]

19. Sheldon, R.A. Characteristic features and biotechnological applications of cross-linked enzyme aggregates (CLEAs). *Appl. Microbiol. Biotechnol.* **2011**, *92*, 467–477. [CrossRef] [PubMed]

20. Mateo, C.; Palomo, J.M.; Fernandez-Lorente, G.; Guisan, J.M.; Fernandez-Lafuente, R. Improvement of enzyme activity, stability and selectivity via immobilization techniques. *Enzyme Microb. Technol.* **2007**, *40*, 1451–1463. [CrossRef]

21. Suescun, A.; Rueda, N.; Dos Santos, J.C.S.; Castillo, J.J.; Ortiz, C.; Torres, R.; Barbosa, O.; Fernandez-Lafuente, R. Immobilization of lipases on glyoxyl-octyl supports: Improved stability and reactivation strategies. *Process Biochem.* **2015**, *50*, 1211–1217. [CrossRef]

22. Peirce, S.; Virgen-Ortíz, J.J.; Tacias-Pascacio, V.G.; Rueda, N.; Barto Lome-Cabrero, R.; Fernandez-Lopez, L.; Russo, M.E.; Marzocchella, A.; Fernandez-Lafuente, R. Development of simple protocols to solve the problems of enzyme coimmobilization. Application to coimmobilize a lipase and a β-galactosidase. *RSC Adv.* **2016**, *6*, 61707–61715. [CrossRef]

23. Marciello, M.; Filice, M.; Palomo, J.M. Different strategies to enhance the activity of lipase catalysts. *Catal. Sci. Technol.* **2012**, *2*, 1531–1543. [CrossRef]

24. Palomo, J.M. Lipases enantioselectivity alteration by immobilization techniques. *Curr. Bioact. Compd.* **2008**, *4*, 126–138. [CrossRef]

25. Filice, M.; Guisan, J.M.; Terreni, M.; Palomo, J.M. Regioselective monodeprotection of peracetylated carbohydrates. *Nat. Protoc.* **2012**, *7*, 1783–1796. [CrossRef] [PubMed]

26. Mateo, C.; Bolivar, J.M.; Godoy, C.A.; Rocha-Martin, J.; Pessela, B.C.; Curiel, J.A.; Munoz, R.; Guisan, J.M.; Fernandez-Lorente, G. Improvement of enzyme properties with a two-step immobilizaton process on novel heterofunctional supports. *Biomacromolecules* **2010**, *11*, 3112–3117. [CrossRef] [PubMed]

27. Henley, J.P.; Sadana, A. Deactivation theory. *Biotechnol. Bioeng.* **1986**, *28*, 1277–1285. [CrossRef] [PubMed]

28. Palomo, J.M.; Segura, R.L.; Mateo, C.; Fernandez-Lafuente, R.; Guisan, J.M. Improving the activity of lipases from thermophilic organisms at mesophilic temperatures for biotechnology applications. *Biomacromolecules* **2004**, *5*, 249–254. [CrossRef] [PubMed]

29. Palomo, J.M.; Fuentes, M.; Fernández-Lorente, G.; Mateo, C.; Guisan, J.M.; Fernández-Lafuente, R. General trend of lipase to self-assemble giving bimolecular aggregates greatly modifies the enzyme functionality. *Biomacromolecules* **2003**, *4*, 1–6. [CrossRef] [PubMed]

30. Ishimoto, T.; Jigawa, K.; Henares, T.G.; Sueyoshi, K.; Endo, T.; Hisamoto, H. Efficient immobilization of the enzyme and substrate for a single-step caspase-3 inhibitor assay using a combinable PDMS capillary sensor array. *RSC Adv.* **2014**, *4*, 7682–7687. [CrossRef]

31. MacArio, A.; Giordano, G.; Frontera, P.; Crea, F.; Setti, L. Hydrolysis of alkyl ester on lipase/silicalite-1 catalyst. *Catal. Lett.* **2008**, *122*, 43–52. [CrossRef]

32. Ghosh, U.; Ganessunker, D.; Sattigeri, V.J.; Carlson, K.E.; Mortensen, D.J.; Katzenellenbogen, B.S.; Katzenellenbogen, J.A. Estrogenic diazenes: Heterocyclic non-steroidal estrogens of unusual structure with selectivity for estrogen receptor subtypes. *Bioorgan. Med. Chem.* **2002**, *11*, 629–657. [CrossRef]

33. Bradford, M.M. A rapid and sensitive method for the quantitation of microgram quantities of protein utilizing the principle of protein-dye binding. *Anal. Biochem.* **1976**, *72*, 248–254. [CrossRef]

34. Laemmli, U.K. Cleavage of Structural Proteins during the Assembly of the Head of Bacteriophage T4. *Nature* **1970**, *227*, 680–685. [CrossRef] [PubMed]

35. Ferna ndez-Lorente, G.; Palomo, J.M.; Mateo, C.; Munilla, R.; Ortiz, C.; Cabrera, Z.; Guisan, J.M.; Fernandez-Lafuente, R. Glutaraldehyde cross-linking of lipases adsorbed on aminated supports in the presence of detergents leads to improved performance. *Biomacromolecules* **2006**, *7*, 2610–2615. [CrossRef] [PubMed]

36. Whistler, R.L.; Wolfrom, M.L.; BeMiller, J.N. *Methods in Carbohydrate Chemistry*; Academic Press: New York, NY, USA; London, UK, 1963; Volume 2.

37. Mateo, C.; Fernandez-Lorente, G.; Pessela, B.C.C.; Vian, A.; Carrascosa, A.V.; Garcia, J.L.; Fernandez-Lafuente, R.; Guisan, J.M. Affinity chromatography of polyhistidine tagged enzymes: New dextran-coated immobilized metal ion affinity chromatography matrices for prevention of undesired multipoint adsorptions. *J. Chromatogr. A* **2001**, *915*, 97–106. [CrossRef]

38. Combes, D.; Yoovidhya, T.; Girbal, E.; Willemot, R.M.; Monsan, P. Mechanism of enzyme stabilization. *Ann. N. Y. Acad. Sci.* **1987**, *501*, 59–62. [CrossRef] [PubMed]

39. Addorisio, V.; Sannino, F.; Mateo, C.; Guisan, J.M. Oxidation of phenyl compounds using strongly stable immobilized-stabilized laccase from Trametes versicolor. *Process Biochem.* **2013**, *48*, 1174–1180. [CrossRef]

*catalysts*

*Article*

# Covalent Immobilization of *Candida rugosa* Lipase at Alkaline pH and Their Application in the Regioselective Deprotection of Per-*O*-acetylated Thymidine

Cintia W. Rivero [†] and Jose M. Palomo *

Departamento de Biocatálisis, Instituto de Catálisis (CSIC), Madrid 28049, Spain; crivero@unq.edu.ar
* Correspondence: josempalomo@icp.csic.es; Tel.: +34-91-585-4768
† Current address: Laboratorio de Investigaciones en Biotecnología Sustentable (LIBioS),
  Universidad Nacional de Quilmes, Roque Sáenz Peña 352, Bernal, Argentina

Academic Editor: Keith Hohn
Received: 10 June 2016; Accepted: 27 July 2016; Published: 2 August 2016

**Abstract:** Lipase from *Candida rugosa* (CRL) was stabilized at alkaline pH to overcome the inactivation problem and was immobilized for the first time by multipoint covalent attachment on different aldehyde-activated matrices. PEG was used as a stabilizing agent on the activity of CRL. At these conditions, CRL maintained 50% activity at pH 10 after 17 h incubation in the presence of 40% ($w/v$) of PEG, whereas the enzyme without additive was instantaneously inactive after incubation at pH 10. Thus, this enzyme was covalently immobilized at alkaline pH on three aldehyde-activated supports: aldehyde-activated Sepharose, aldehyde-activated Lewatit105 and heterofunctional aldehyde-activated EDA-Sepharose in high overall yields. Heterogeneous stable CRL catalysts at high temperature and solvent were obtained. The aldehyde-activated Sepharose-CRL preparation maintained 70% activity at 50 °C or 30% ($v/v$) acetonitrile after 22 h and exhibited high regioselectivity in the deprotection process of per-*O*-acetylated thymidine, producing the 3′-OH-5′-OAc-thymidine in 91% yield at pH 5.

**Keywords:** *Candida rugosa* lipase; stabilization; covalent immobilization; PEG; alkaline pH; regioselectivity; nucleosides

---

## 1. Introduction

Enzymes are natural catalysts working at mild conditions with high specificity by substrate and excellent selectivity which could constitute a green solution for the industry. However, enzymes are unstable out of their natural environments, and parameters such as pH and temperature (T) are relevant to their stability. Therefore, the stabilization mechanism represents an important issue in enzyme development for possible industrial implementation. Low enzyme stability has been overcome using several strategies: genetic engineering [1], chemical modifications [2], the addition of stabilizing agents [3–6], or the use of different immobilization mechanisms [7,8].

Particularly, multipoint covalent attachment on macroporous supports has been described as a very interesting approach to stabilize enzymes [9–11].

The functionalization of the support materials by molecules as short spacer arms assures the attachment of an enzyme molecule to the matrix through various covalent linkages. The amino-acidic residues of the protein involved in the covalent immobilization should be rigid, conserving the relative positions against changes in the protein conformation, e.g., with the effect of distorting agents, such as heat and organic solvents [12–14].

In this way the use of supports functionalized by aldehyde groups under alkaline conditions promotes enzyme immobilization by a specific orientation, through the richest area containing the highest number of lysines, generating a multipoint covalent interaction [15]. The advantages of the aldehyde groups are: (i) they are reactive toward unprotonated primary amines; (ii) they are stable under alkaline conditions; (iii) they do not exhibit steric hindrances for intramolecular reactions [16]. In fact, several industrial enzymes have been stabilized by multipoint covalent attachment on these aldehyde-activated supports [15].

However, enzymes sensitive to alkaline pH are not able to be immobilized through this methodology. *Candida rugosa* lipase (CRL), a very useful enzyme in biotransformations [17–19], is one of these enzymes with very limited conditions for proper use, especially in regards to pH [20]. The immobilization methods for the covalent attachment of this lipase described in the literature use the application of functionalized supports, e.g., with glutaraldehyde or cyanogen bromide groups, for covalent immobilization at neutral pH, which mainly correspond to the reaction to the terminal amino group, with a low, intense covalent immobilization [20,21]. Also, the adsorption on hydrophobic supports at neutral pH has been used for improving its stability, although this is a reversible immobilization and the enzyme can be leached from the support in the presence of some concentration of solvent or detergent [20]. Therefore, the multipoint covalent immobilization of this enzyme on aldehyde supports would be an excellent strategy, considering the advantages described, for obtaining an irreversible, immobilized biocatalyst with high stability.

Improvements of the enzyme's stability with the presence of some additives such as polyols, solvents or sugars have been reported [22,23].

The use of these additives has the advantage of providing enzyme activity at high temperatures and alkaline pH [24]. A study of the protective effect of PEG, trehalose and glycerol revealed an increase with the reagent concentration and length of the carbon chain [25,26]. These compounds are known to have a more or less pronounced effect on water activity and on the degree of water molecule association [25]. On the other hand, aqueous solutions of polyols, polymers and sugars, as additives, were used to study the thermostabilization of enzymes [27].

Herein, we propose a methodology to stabilize *Candida rugosa* lipase (CRL) at alkaline pH using PEG as an additive, permitting for the first time its immobilization by multipoint covalent attachment on aldehyde-activated (Ald) derivative supports. Two different covalent immobilization strategies involving different orientations of the protein (Figure 1) in the immobilization were used (Scheme 1). These new immobilized CRL biocatalysts were used to catalyze the regioselective monodeprotection of per-*O*-acetylated thymidine, an interesting intermediate in the synthesis of different fungicidal, antitumor, and especially antiviral agents [28].

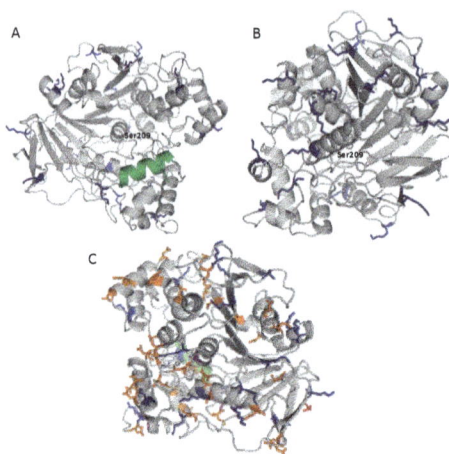

**Figure 1.** Structure of mature CRL. (**A**) Lid side. Oligopeptide lid (green), Lys (blue); (**B**) Lid opposite side marked lysines; (**C**) Lid opposite side marked lysines and aspartic and alutamic acids (orange). The structure of CRL was obtained from the Protein Data Bank (pdb code: CRL) and the picture was created using Pymol v. 0.99.

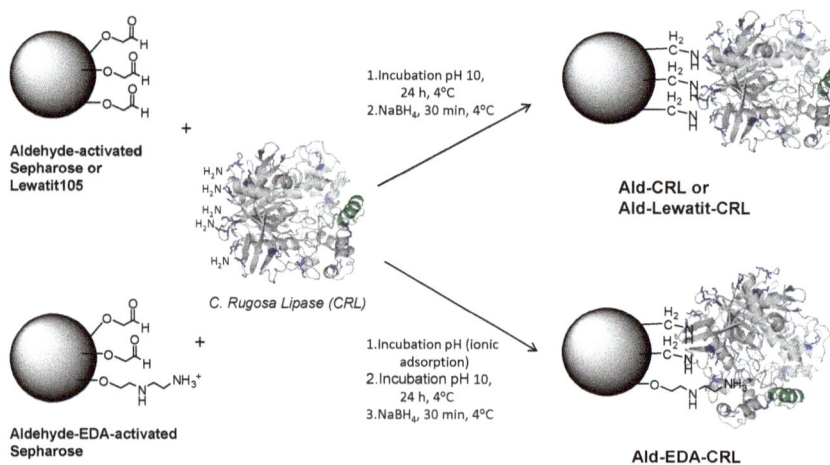

**Scheme 1.** Covalent immobilization of CRL by different methodologies.

## 2. Results and Discussions

### 2.1. Stabilization of CRL at Alkaline pH

The stability of soluble pure CRL by incubation at different pHs was first studied (Figure 2). This lipase has been described to be a very sensitive enzyme [20]. After purification, the enzyme incubated at pH 7 maintained around 92% activity during 25 h at 4 °C. When the pH was increased up to 9, 60% activity was found after 25 h. However, the activity of the enzyme was totally lost immediately after incubation at pH 10, exhibiting the extremely low stability at alkaline pH, which is mandatory to perform the multipoint covalent immobilization on an aldehyde-activated support.

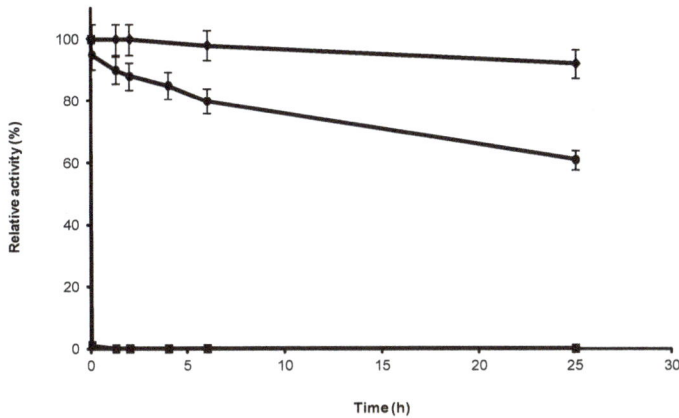

**Figure 2.** Stability of purified CRL at 25 °C and different pHs: pH 7 (rhombus), pH 9 (circles), pH 10 (squares); 0.27 mg purified free lipase was used in each experiment.

Thus, different additives—at 20% $(w/v)$ concentration—were added to the enzyme solution to study their effect on the enzyme stability (Figure 3). The best result was achieved when PEG was previously added to the lipase alkaline solution (pH 10), where the enzyme retained 40% activity after incubation for 4.5 h, whereas the soluble enzyme without additives maintained only 20% activity. Using polyols such as glycerol, dextran or trehalose, a slightly improved activity was observed.

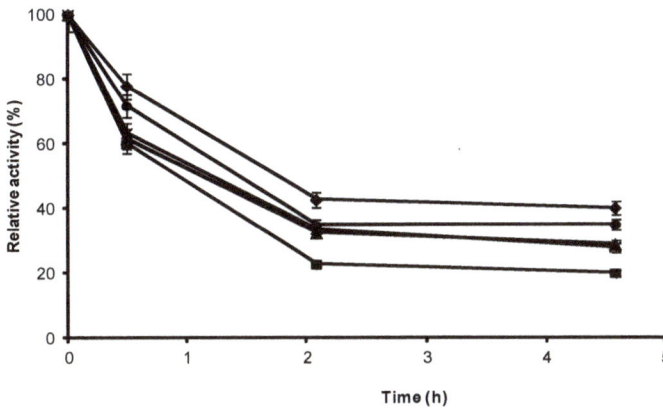

**Figure 3.** Stabilization of CRL in the presence of different additives. The additives were added at 20% $(w/v)$ concentration and the experiments were performed at pH 10 and 4 °C. Without additive (squares), PEG (rhombus), glycerol (circles), dextran (triangles), threalose (×).

PEG was selected as a stabilizing agent and the effect of the additive concentration and the molecular size on the lipase stability was studied (Figure 4). PEG1500 stabilized the enzyme slightly better than PEG6000 at a 15% concentration $(w/v)$, although the best results were found using 40% $(w/v)$ PEG1500, where CRL conserved 50% activity after 17 h incubation (Figure 4).

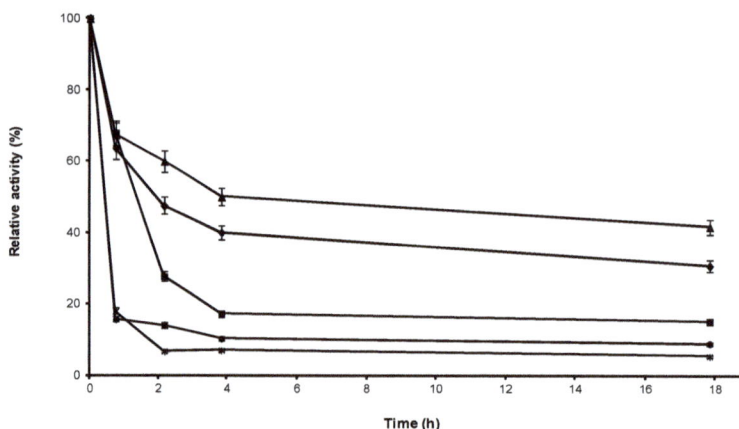

**Figure 4.** Effect of PEG concentration on the stability of CRL at 4 °C and pH 10. Without additive (×), PEG-1500 15% (squares), PEG-6000 15% (circles), PEG-1500 20% (rhombus), PEG-1500 40% (triangles).

The possible mechanism for the high stabilization of lipase achieved by the addition of PEG could be explained by two different effects. The first is (1) a strong physical adsorption of the PEG molecules to the hydrophobic area of the protein, as previously has been reported [29]. Especially here, in lipases the most hydrophobic area is concentrated on the lid and the surrounding active site; therefore, the PEG could generate protection of the active site, significantly improving the stability in an extreme condition such as alkaline pHs. The second effect is (2) due to the fact that the use of higher PEG concentrations causes a high viscosity which also may prevent undesired changes in the enzyme structure promoted by the alkaline pH [30].

### 2.2. Multipoint Covalent Immobilization of CRL

The enzyme was immobilized at optimal conditions (with 40% PEG in solution) using two different strategies to get multipoint covalent immobilization of the lipase throughout different orientations (Scheme 1). CRL was immobilized on Ald-Sepharose at 89% yield (loading of 4.9 mg lipase/g support) at pH 10.2 over 24 h retaining 53% initial activity, whereas a 69% yield of lipase immobilized maintaining 39% of initial activity was obtained using Ald-Lew105 (Table 1).

**Table 1.** Covalent immobilization of *Candida rugosa* lipase on different supports at 4 °C.

| Support | Immobilization yield (%) [a] | Retained Activity (%) |
|---|---|---|
| Ald-Sepharose | 89 | 53 |
| Ald-Lew105 | 69 | 39 |
| Ald-EDA-Sepharose | 95 | 37 |

[a] Immobilization for 24 h; 5.5 mg pure lipase was offered per gram of support.

The immobilization on heterofunctional Ald-EDA-Sepharose was performed at pH 8 and around 90% yield was achieved after 3 h incubation. After that, the immobilized preparation was incubated at pH 10 for 24 h to promote a possible multipoint covalent attachment. The immobilized preparation retained 37% overall initial activity (Table 1).

### 2.3. Stability of Different Covalent Immobilized Preparations of CRL

To evaluate the effect of the immobilization method on the stabilization of the enzyme, inactivation experiments at different conditions were studied (Figures 5 and 6).

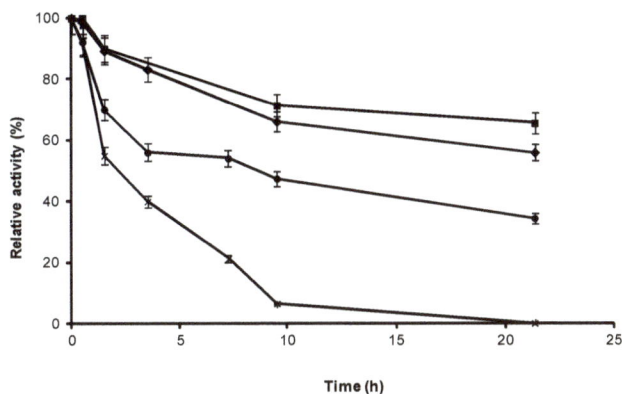

**Figure 5.** Thermal inactivation course of different CRL immobilized preparations. Experiments were performed at 50 °C and pH 5. Free CRL (×), Ald-EDA-CRL (rhombus), Ald-CRL (squares), Ald-Lew105-CRL (circles).

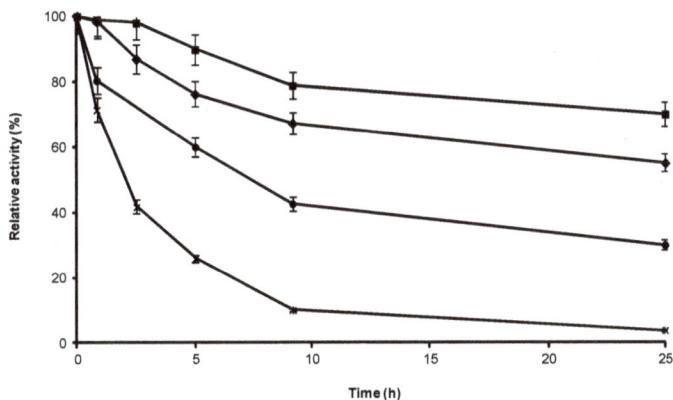

**Figure 6.** Inactivation profile of different CRL immobilized preparations in the presence of co-solvent. Experiments were carried out at 25 °C, pH 5 and 30% ($v/v$) acetonitrile. Free CRL (×), Ald-EDA-CRL (rhombus), Ald-CRL (squares), Ald-Lew105-CRL (circles).

At 50 °C, the best stabilization of CRL was achieved after immobilization on Ald-Sepharose. The Ald-CRL, Ald-lew105-CRL and Ald-EDA-CRL immobilized preparations conserved more than 50% activity after 9 h incubation at 50 °C, whereas the free lipase only retained 6% activity (Figure 5). Indeed, the Ald-CRL preparation still conserved 70% of the initial activity after 22 h at 50 °C whereas the Ald-Lew105-CRL preparation only maintained 38% activity at this time. This demonstrates the effect of the matrix on the lipase stabilization.

Also, of the differently oriented catalysts, Ald-EDA-CRL showed good activity, maintaining around 60% activity after 22 h incubation (Figure 5).

When the CRL preparations were incubated at 30% ($v/v$) acetonitrile at 25 °C, the effect was even clearer (Figure 6). The Ald-CRL preparation was the most stable catalyst also in the presence of a co-solvent, retaining 70% activity after 25 h incubation, conditions where the soluble enzyme was completely inactive. The enzyme immobilized on the Ald-EDA support conserved 55% activity whereas the Ald-Lew105-CRL preparation was again less stable than the Sepharose one (Figure 6).

## 2.4. Regioselective Deprotection of Per-O-acetylated Thymidine by Immobilized CRL Biocatalysts

The different covalent immobilized preparations of CRL were used as catalysts in the hydrolytic deacetylation of per-*O*-acetylated thymidine (**1**) at different pHs (Table 2).

**Table 2.** Regioselective deprotection of 3,5-*O*-diacetylated thymidine **1** with different aldehyde-activated CRL biocatalysts at 25 °C in aqueous media.

| Biocatalyst | pH | Initial Rate [a] | Reaction Time (h) | Yield [b] (%) | 2 (%) | 3 (%) | Thymidine |
|---|---|---|---|---|---|---|---|
| free CRL | 5.0 | 0.08 | 104 | 99 | 81 | 9 | 10 |
| Ald-CRL | 5.0 | 0.24 | 71 | 100 | 91 | 3 | 6 |
| Ald-EDA-CRL | 5.0 | 0.09 | 144 | 100 | 70 | 10 | 20 |
| Lew105-CRL | 5.0 | 0.06 | 144 | 100 | 17 | 15 | 67 |
| free CRL | 7.0 | 0.08 | 104 | 99 | 90 | 8 | 2 |
| Ald-CRL | 7.0 | 0.26 | 51 | 100 | 90 | 4 | 6 |
| Ald-EDA | 7.0 | 0.09 | 152 | 100 | 88 | 10 | 3 |
| Lew105 | 7.0 | 0.04 | 150 | 62 | 28 | 32 | 2 |
| free CRL | 8.0 | 0.08 | 120 | 100 | 75 | 11 | 13 |
| Ald-CRL | 8.0 | 0.11 | 73 | 100 | 88 | 8 | 4 |

[a] the initial rate in $\mu mol \times mgprot^{-1} \times h^{-1}$. It was calculated at 10%–30% conversion. [b] yield of the monohydroxy acetylated product at 100% conversion.

The Ald-CRL preparation showed the highest activity (three times higher than the soluble enzyme) and regioselectivity in the monoacetylation of **1** at pH 5 and 7, producing the C-3 hydroxy monoacetylated thymidine **2** in around 90% yield. The other CRL preparations showed lower regioselectivity, and no differences in activity compared with the soluble enzyme (Table 2). In particular, CRL immobilized on Lew105 lost the specificity and the regioselectivity. Also, the results using the Ald-CRL preparation at pH 8 were better than using the soluble enzyme (Table 2). The role of the immobilization method on the modulation of the activity and regioselectivity has been shown. In the case of CRL, immobilization by this strategy generates a particular orientation of the enzyme (from the protein lid's opposite side (Figure 1C)) and a strong rigidification of its structure. This phenomenon alters the open-closed movement of the oligopeptide lid during catalysis and the exact shape of the open structure which is translated in a significant modulation of the lipase properties. We have already observed these alterations in lipase enantioselectivity [31], and also in the regioselective deprotection of different glycoderivatives [18], and it has been recently observed in the production of 2-glyceryl derivatives [32].

## 3. Experimental Section

### 3.1. Materials

Lipase from *Candida rugosa* (CRL), ethylendiamine (EDA), p-nitrophenyl butyrate (pNPB), dithiothreitol (DTT), polyethyleneglycol (PEG) (Mr 1500, 6000), glycerol, dextran (Mw 1500) and trehalose were from Sigma Chem. Co (St. Louis, MO, USA). Sepharose 10BCL, octyl-Sepharose and cyanogen bromide (CNBr) activated Sepharose beads were from GE-Healthcare (Uppsala, Sweden).

Aldehyde-activated Sepharose or Lewatit VP OC105 (Ald or Ald-Lew105) were prepared as previously described [33]. 3,5-*O*-diacetylated thymidine **1** was prepared as previously described [34].

### 3.2. Lipase Activity Assay

The activities of the soluble lipase (without additives or in the presence of different concentrations of PEG, glycerol, DTT, trehalose or dextran), supernatant and enzyme suspension were analyzed spectrophotometrically measuring the increment in absorbance at 348 nm produced by the release of p-nitrophenol (pNP) ($\varepsilon$= 5.150 $M^{-1} \cdot cm^{-1}$) in the hydrolysis of 0.4 mM pNPB in 25 mM sodium phosphate at pH 7 and 25 °C. To initialize the reaction, 0.05–0.2 mL of lipase solution or suspension was added to 2.5 mL of substrate solution in magnetic stirring. Enzymatic activity is given as one µmol of p-nitrophenol released per minute per mg of enzyme (IU) under the conditions described above.

### 3.3. C. rugosa Lipase Purification

The enzyme was purified from commercial crude extract by interfacial adsorption as previously described [35]. Lipase commercial extract was dissolved in 25 mM sodium phosphate buffer at pH 7 to give 200 mg extract/mL, and submitted to gentle stirring during 1 h at 4 °C, and centrifuged at 12,000 rpm during 15 min. The supernatant was separated from the pellet and the protein amount was calculated by Bradford method [36] (5.5 mg prot/mL). Then 1 mL of this supernatant was diluted in 9 mL of 25 mM phosphate buffer pH 7 and the solution was added to one gram of octyl-Sepharose. The reaction was performed at 4 °C for 1 h. After that, the suspension was filtered by vacuum and the solid was washed several times with distilled water. More than 95% of the enzyme was immobilized.

For the preparation of the covalent immobilized catalysts, the lipase was desorbed from the support (one gram of octyl-CRL) adding 20 mL of a solution of 25 mM phosphate buffer pH 7 with 0.4% Triton X-100 (*v*/*v*) and incubated it for 1 h. A final solution of 0.27 mg purified lipase/mL was obtained.

### 3.4. Preparation of EDA-Aldehyde–Activated Sepharose Support (Ald-EDA)

Sepharose 10 BCL (10 g) was suspended in a mixture solution of 44 mL water, 16 mL acetone, 3.28 g NaOH, 0.2 g NaBH4 and 11 mL epichlorhydrine. The suspension was stirred mildly for 16 h and washed with an excess of water. One gram of epoxy-Sepharose support was suspended in 10 mL of ethylenediamine (0.1 M) solution at pH 8 for 6 h. Finally the support was oxidized adding a solution of 10 mL of water with 140 µmol of sodium periodate per gram of support during 90 minutes and washed abundantly with distilled water and store at 4 °C.

### 3.5. Multipoint Covalent Immobilization of CRL on Different Aldehyde-Activated Sepharose Supports (Ald)

#### 3.5.1. Immobilization on Aldehyde-Activated Sepharose (Ald) or Aldehyde-Activated Lewatit-105 (Ald-Lew105) at Alkaline pH

First 20 mL of lipase solution (0.27 mg$_{lipase}$/mL) was dissolved in 20 mL solution of 100 mM sodium bicarbonate pH 8.2 containing 40% PEG1500 (*w*/*v*) and the pH was adjusted at pH 10.15. After that, one gram of Ald-Sepharose or Ald-Lew-105 was added and the reaction was maintained during 16 h at 4 °C. Finally the enzyme-support multi-interaction was ended by adding 1 mg of sodium borohydride per mL of suspension during 30 min [33] (Scheme 1). The immobilization yields are shown in Table 1.

#### 3.5.2. Immobilization on Aldehyde-Activated EDA-Sepharose (Ald-EDA) at pH 8 and Incubation at pH 10

One gram of Ald-EDA-Sepharose was added to 10 mL of purified CRL solution (0.27 mg lip/mL) containing 40% PEG1500 (*w*/*v*). Then the suspension was stirred for 2 h at pH 8 and 4 °C. Periodically, samples of the supernatants and suspensions were withdrawn, and the enzyme activity was measured as described above. After the preparation was filtrated by vacuum and the solid was incubated in 10 mL sodium bicarbonate buffer at pH 10 for 24 h. Finally, the preparation was reduced by addition of 10 mg sodium borohydride for 30 minutes and then washed with water (Scheme 1). The immobilization yields are shown in Table 1.

*3.6. Inactivation of CRL Immobilized Preparations against T and Co-Solvent*

First 0.5 g of biocatalyst were dissolved in 5 mL of 25 mM sodium phosphate buffer (with 30% ($v/v$) acetonitrile) at 25 °C or incubated at 50 °C in acetate buffer at pH 5. The remaining activity at different times was measured by the assay described above using pNPB as substrate.

*3.7. Enzymatic Hydrolysis of 3′,5′-Di-O-Acetylthymidine (**1**)*

Substrate **1** (5 mM) was dissolved in a mixture of acetonitrile (5%, $v/v$) in 10 mM sodium phosphate at pH 7.0 or 10 mM sodium acetate at pH 5.0. 0.2 g of biocatalyst was added to 2 mL of this solution at 25 °C. During the reaction, the temperature and the pH value was maintained constant using a pH-stat Mettler Toledo DL50 graphic (Mettler-Toledo, LLC 1900 Polaris Parkway, Columbus, OH, USA). The degree of hydrolysis was analyzed by reverse phase HPLC (Spectra Physic Thermo SP 100 coupled with an UV detector Spectra Physic SP 8450 (Thermo Fisher-Scientific, Waltham, MA, USA). For these assays a Kromasil C18 5 μm φ (25 cm×0.4 cm) column was used and the following gradient program (A: mixture of acetonitrile (10%, $v/v$) in 10mM ammonium phosphate at pH 4.2; B: mixture of miliQ water (10%, $v/v$) in acetonitrile; method: 0–6 min 100% A, 6–14 min 85% A to 15%B, 14–22 min 100% A, flow: 1.0 mL·min$^{-1}$). UV detection was performed at 260 nm. The unit of enzymatic activity was defined as micromoles of substrate hydrolyzed per minute per mg of immobilized protein. The monodeprotected 3-OH (**2**) and 5-OH (**3**) were used as pure standards. The retention time was 2.4 min for Thymidine, 9.4 min for **2** and 10.2 min 5 and 19 min for **1**.

**4. Conclusions**

Lipase from *C. rugosa* has been stabilized at alkaline pH to overcome the inactivation problem by the addition of PEG1500 as a stabilizing agent. Therefore, this has permitted its immobilization for the first time by multipoint covalent attachment on different aldehyde-activated supports in high overall yields. Very stable CRL biocatalysts have been prepared; in particular CRL immobilized on Ald-Sepharose was much more stable than the soluble enzyme. This stable biocatalyst showed an excellent regioselectivity in the monodeprotection of per-O-acetylated thymidine, producing the 3-OH-5′-OAc-thymidine in 91% yield at pH 5. Therefore, this new biocatalyst represents an interesting alternative to the octyl-CRL preparation, which has been described as an excellent catalyst in nucleosides and especially in monosaccharides deprotection [18]. However, Ald-CRL presents the advantage of being an irreversible catalyst with high stability in the presence of solvent. This strategy can be also extended to other pH-sensitive enzymes to generate highly stable and active biocatalysts.

**Acknowledgments:** This work has been sponsored by CSIC. The authors gratefully recognize the support from National Scientific and Technical Research Council (CONICET).

**Author Contributions:** J.M.P. conceived and designed the experiments; C.W.R. performed the experiments; J.M.P. wrote the paper.

**Conflicts of Interest:** The authors declare no conflict of interest.

**References**

1. Reetz, M.T. Biocatalysis in Organic Chemistry and Biotechnology: Past, Present, and Future. *J. Am. Chem. Soc.* **2013**, *135*, 12480–12496. [CrossRef] [PubMed]
2. Xue, Y.; Wu, C.-Y.; Branford-White, C.J.; Ning, X.; Nie, H.-L.; Zhu, L.-M. Chemical modification of stem bromelain with anhydride groups to enhance its stability and catalytic activity. *J. Mol. Catal. B Enzym.* **2010**, *63*, 188–193. [CrossRef]

3. Srirangsan, P.; Kawai, K.; Hamada-Sato, N.; Watanabe, M.; Suzuki, T. Stabilizing effects of sucrose-polymer formulations on the activities of freeze-dried enzyme mixtures of alkaline phosphatase, nucleoside phosphorylase and xanthine oxidase. *Food Chem.* **2011**, *125*, 1188–1193. [CrossRef]

4. Santagapita, P.R.; Brizuela, L.G.; Mazzobre, M.F.; Ramírez, H.L.; Corti, H.R.; Santana, R.V.; Buera, M.P. β-cyclodextrin modifications as related to enzyme stability in dehydrated systems: Supramolecular transitions and molecular interactions. *Carbohydr. Polym.* **2011**, *83*, 203–209. [CrossRef]

5. Nascimento, C.; Leandro, J.; Lino, P.R.; Ramos, L.; Almeida, A.J.; De Almeida, I.T.; Leandro, P. Polyol additives modulate the in vitro stability and activity of recombinant human phenylalanine hydroxylase. *Appl. Biochem. Biotechnol.* **2010**, *162*, 192–207. [CrossRef] [PubMed]

6. Iyer, P.V.; Ananthanarayan, L. Enzyme stability and stabilization-Aqueous and non-aqueous environment. *Process Biochem.* **2008**, *43*, 1019–1032. [CrossRef]

7. Guisán, J.M. *Immobilization of Enzymes and Cells, Methods in Molecular Biology*, 3rd ed.; Humana Press: New York, NY, USA, 2013; pp. 1–377.

8. Mateo, C.; Grazu, V.; Palomo, J.M.; Lopez-Gallego, F.; Fernandez-Lafuente, R.; Guisán, J.M. Immobilization of Enzymes on Heterofunctional Epoxy Supports. *Nat. Protoc.* **2007**, *2*, 1022–1033. [CrossRef] [PubMed]

9. Katchalski-Katzir, E. Immobilized enzymes—Learning from past successes and failures. *Trends Biotechnol.* **1993**, *11*, 471–478. [CrossRef]

10. Cao, L. Immobilised enzymes: Science or art? *Curr. Opin. Chem. Biol.* **2005**, *9*, 217–226. [CrossRef] [PubMed]

11. Mateo, C.; Palomo, J.M.; Fernandez-Lorente, G.; Guisan, J.M.; Fernandez-Lafuente, R. Improvement of enzyme activity, stability and selectivity via immobilization techniques. *Enzyme Microb. Technol.* **2007**, *40*, 1451–1463. [CrossRef]

12. Pedroche, J.; Yust, M.M.; Mateo, C.; Fernández-Lafuente, R.; Girón-Calle, J.; Alaiz, M.; Vioque, J.; Guisan, J.M.; Millan, F. Effect of the support and experimental conditions in the intensity of the multipoint covalent attachment of proteins on aldehyde activated-Sepharose supports: Correlation between enzyme–support linkages and thermal stability. *Enzyme Microb. Technol.* **2007**, *40*, 1160–1167. [CrossRef]

13. Bolivar, J.M.; Rocha-Martin, J.; Mateo, C.; Cava, F.; Berenguer, J.; Vega, D.; Fernandez-Lafuente, R.; Guisan, J.M. Purification and stabilization of a glutamate dehygrogenase from Thermus thermophilus via oriented multisubunit plus multipoint covalent immobilization. *J. Mol. Catal. B Enzym.* **2009**, *58*, 158–163. [CrossRef]

14. Klibanov, A.M. Stabilization of Enzymes against Thermal Inactivation. *Adv. Appl. Microbiol.* **1983**, *29*, 1–28. [PubMed]

15. Mateo, C.; Abian, O.; Bernedo, M.; Cuenca, E.; Fuentes, M.; Fernandez-Lorente, G.; Palomo, J.M.; Grazu, V.; Pessela, B.C.C.; Giacomini, C.; et al. Some special features of aldehyde activated supports to immobilize proteins. *Enzyme Microb. Technol.* **2005**, *37*, 456–462. [CrossRef]

16. Blanco, R.M.; Calvete, J.J.; Guisan, J.M. Immobilization-stabilization of enzymes; variables that control the intensity of the trypsin (amine)-Sepharose (aldehyde) multipoint attachment. *Enzyme Microb. Technol.* **1989**, *11*, 353–359. [CrossRef]

17. Kolodiazhnyi, O.I. Recent developments in the asymmetric synthesis of -chiral phosphorus compounds. *Tetrahedron Asymmetry* **2012**, *23*, 1–46.

18. Filice, M.; Guisan, J.M.; Terreni, M.; Palomo, J.M. Regioselective monodeprotection of peracetylated carbohydrates. *Nat. Protoc.* **2012**, *7*, 1783–1796. [CrossRef] [PubMed]

19. Domínguez de María, P.; Alcántara, A.R.; Carballeira, J.D.; de la Casa, R.M.; García-Burgos, C.A.; Hernáiz, M.J.; Sánchez-Montero, J.M.; Sinisterra, J.V. Candida rugosa Lipase: A traditional and complex biocatalyst. *Curr. Org. Chem.* **2006**, *10*, 1053–1066. [CrossRef]

20. Palomo, J.M.; Fernández-Lorente, G.; Mateo, C.; Ortiz, C.; Fernandez-Lafuente, R.; Guisan, J.M. Modulation of the enantioselectivity of lipases via controlled immobilization and medium engineering Hydrolytic resolution of Mandelic acid esters. *Enzyme Microb. Technol.* **2002**, *31*, 775–783. [CrossRef]

21. Knezevic, Z.; Milosavic, N.; Bezbradica, D.; Jakovljevic, Z.; Prodanovic, R. Immobilization of lipase from Candida rugosa on Eupergit® C supports by covalent attachment. *Biochem. Eng. J.* **2006**, *30*, 269–278. [CrossRef]

22. Noriko, M.; Miok, K.; Tadao, K. Stabilization of L-Ascorbic acid by superoxide dismutase and catalase. *Biosci. Biotechnol. Biochem.* **1999**, *63*, 54–57.

23. Lozano, P.; Combes, D.; Iborra, J.L. Effects of polyols on chymotrypsin thermostability a mechanistic analysis of the enzyme stabilization. *J. Biotechnol.* **1994**, *35*, 9–18. [CrossRef]

24. Castro, G.R. Properties of soluble α-chymotrypsin in neat glycerol and water. *Enzyme Microb. Technol.* **2000**, *27*, 143–150. [CrossRef]

25. Combes, D.; Yoodikhya, T.; Girbal, E.; Willemot, R.M.; Monsan, P. Mechanism of Enzyme Stabilization. *Ann. N. Y. Acad. Sci.* **1987**, *501*, 59–62. [CrossRef] [PubMed]

26. Salahas, G.; Peslis, B.; Georgiou, C.D.; Gavalas, N.A. Trehalose, an extreme temperature protector of phosphoenolpyruvate carboxylase from the C4-plant Cynodon dactylon. *Phytochemistry* **1997**, *46*, 1331–1334. [CrossRef]

27. Breccia, J.D.; Moran, A.C.; Castro, G.R.; Siñeriz, F. Thermal stabilization by polyols of β-xylanase from Bacillus amyloliquefaciens. *J. Chem. Technol. Biotechnol.* **1998**, *71*, 241–245. [CrossRef]

28. Jordheim, L.P.; Durantel, D.; Zoulim, F.; Dumontet, C. Advances in the development of nucleoside and nucleotide analogues for cancer and viral diseases. *Nat. Rev. Drug Discov.* **2013**, *12*, 447–464. [CrossRef] [PubMed]

29. Rawat, S.; Raman Suri, C.; Sahoo, D.K. Molecular mechanism of polyethylene glycol mediated stabilization of protein. *Biochem. Biophys. Res. Commun.* **2010**, *392*, 561–566. [CrossRef] [PubMed]

30. Fernández-Lorente, G.; Lopez-Gallego, F.; Bolivar, J.M.; Rocha-Martin, J.; Moreno-Perez, S.; Guisan, J.M. Immobilization of proteins on highly activated glyoxyl supports: Dramatic increase of the enzyme stability via multipoint immobilization on pre-existing carriers. *Curr. Org. Chem.* **2015**, *19*, 1719–1731. [CrossRef]

31. Palomo, J.M. Lipases Enantioselectivity Alteration by Immobilization Techniques. *Curr. Biol. Compet.* **2008**, *4*, 126–138. [CrossRef]

32. Guajardo, N.; Bernal, C.; Wilson, L.; Cabrera, Z. Selectivity of R-α-monobenzoate glycerol synthesis catalyzed by Candida antarctica lipase B immobilized on heterofunctional supports. *Process Biochem.* **2015**, *50*, 1870–1877. [CrossRef]

33. Mateo, C.; Palomo, J.M.; Fuentes, M.; Betancor, L.; Grazu, V.; Lopez-Gallego, F.; Pessela, B.C.C.; Hidalgo, A.; Fernandez-lorente, G.; Fernandez-lafuente, R.; et al. Glyoxyl agarose: A fully inert and hydrophilic support for immobilization and high stabilization of proteins. *Enzyme Microb. Technol.* **2006**, *39*, 274–280. [CrossRef]

34. Romero, O.; Filice, M.; de Las Rivas, B.; Carrasco-Lopez, C.; Klett, J.; Morreale, A.; Hermoso, J.A.; Guisan, J.M.; Abian, O.; Palomo, J.M. Semisynthetic peptide-lipase conjugates for improved biotransformations. *Chem. Commun.* **2012**, *48*, 9053–9055. [CrossRef] [PubMed]

35. Bastida, A.; Sabuquillo, P.; Armisen, P.; Fernandez-Lafuente, R.; Huguet, J.; Guisan, J.M. A single step purification, immobilization, and hyperactivation of lipases via interfacial adsorption on strongly hydrophobic supports. *Biotechnol. Bioeng.* **1998**, *58*, 486–493. [CrossRef]

36. Bradford, M.M. A rapid and sensitive method for the quantitation of microgramquantities of protein utilizing the principle of protein dye binding. *Anal. Biochem.* **1976**, *72*, 248–254. [CrossRef]

![catalysts logo] *catalysts*

MDPI

*Review*

# Tandem Reactions Combining Biocatalysts and Chemical Catalysts for Asymmetric Synthesis

**Yajie Wang and Huimin Zhao \***

Department of Chemical and Biomolecular Engineering, University of Illinois at Urbana-Champaign, Urbana, IL 61801, USA; ywang345@illinois.edu
\* Correspondence: zhao5@illinois.edu; Tel.: +1-217-333-2631; Fax: +1-217-333-5052

Academic Editors: Jose M. Palomo and Cesar Mateo
Received: 7 November 2016; Accepted: 29 November 2016; Published: 5 December 2016

**Abstract:** The application of biocatalysts in the synthesis of fine chemicals and medicinal compounds has grown significantly in recent years. Particularly, there is a growing interest in the development of one-pot tandem catalytic systems combining the reactivity of a chemical catalyst with the selectivity engendered by the active site of an enzyme. Such tandem catalytic systems can achieve levels of chemo-, regio-, and stereo-selectivities that are unattainable with a small molecule catalyst. In addition, artificial metalloenzymes widen the range of reactivities and catalyzed reactions that are potentially employable. This review highlights some of the recent examples in the past three years that combined transition metal catalysis with enzymatic catalysis. This field is still in its infancy. However, with recent advances in protein engineering, catalyst synthesis, artificial metalloenzymes and supramolecular assembly, there is great potential to develop more sophisticated tandem chemoenzymatic processes for the synthesis of structurally complex chemicals.

**Keywords:** tandem catalysis; chemoenzymatic; biocatalysis; dynamic kinetic resolution; artificial metalloenzyme; asymmetric synthesis

---

## 1. Introduction

Living beings do not use enzymes in isolation. However, they build up the living system by applying multi-step synthesis strategies catalyzed by enzymes acting cooperatively. In that way, complex molecules are built from simple elements through multi-step biosynthetic routes. The cooperative action of a sequence of enzymatic reactions unveils the mysteries of "perfect" reaction systems with maximized energy utilization efficiency and minimal waste generation. Such synergy inspires biochemists to mimic nature to develop multi-step catalysis for selective synthesis, termed tandem catalysis. Compared with stepwise synthesis, one-pot tandem reactions offer an attractive approach to improve the overall synthetic efficiency by eliminating the purification steps of intermediates, suppressing the side reactions and enhancing selectivity of the product by building dynamic equilibrium in each step (which improves productivity by allowing equilibrium reactions to proceed to nearly full conversion). Thus, it is not surprising that from synthetic and industrial standpoints, there is an increasing interest in eco-friendly tandem processes.

Tandem catalysis employing the same type of catalyst, such as multi-step chemocatalysis or biocatalysis, has been extensively studied [1,2]. Recently, there is a new trend to combine chemocatalysis and biocatalysis in one-pot to obtain synergistic synthetic abilities that cannot be achieved by either separately [3]. Chemocatalysis and biocatalysis are generally considered two different fields, each with their unique considerations. The catalysts from these fields either catalyze completely different reactions, or similar reactions with different rates, selectivity and substrate scopes. Importantly, the catalysts from each of these fields have distinct advantages and limitations. Organometallic catalysts play a key role in the manufacturing of chemicals. They have wide substrate scopes and high productivity, but they usually show poor regio-, stereo- and enantio-selectivity that have to be overcome by tedious ligand design. Most of the reactions catalyzed by transition-metal complexes are performed under harsh conditions, such as high temperature and pressure. Biocatalysis is becoming more widely used in the pharmaceutical industry due to significant advances in enzyme discovery, supply and improvements. There has also been an increase in applications of these biocatalysts for chiral catalysis and green chemistry [4–7]. Biocatalysts typically have high regio-, stereo- and enantio-selectivity, but low productivity. Considering those factors, combining the two technologies into a tandem one-pot reaction would allow access to more enantiopure compounds.

Unlike developing tandem catalysis with the same type of catalyst, it is more challenging to combine chemocatalysis and biocatalysis in one-pot due to mutual inactivation. With the exception of lipases and serine proteases, the majority of enzymes are not able to maintain high catalytic activity in organic solvents and at high temperatures. Similarly, most transition-metal complexes are inhibited in aqueous solution with or without cellular components. Several strategies have been developed to overcome these obstacles, including using biphasic systems, the development of supramolecular hosts and the development of artificial metalloenzymes to compartmentalize the chemical catalysts. Both metal catalysts and biocatalysts have been engineered to show higher activity in aqueous solutions or organic solvents with the utilization of catalyst immobilization and protein engineering respectively. In this review, we will cover major accomplishments in one-pot chemoenzymatic reactions within the last three years, including dynamic kinetic resolution, one-pot concurrent transformations in aqueous solutions, and interfacing transition-metal complexes with living cells.

## 2. Dynamic Kinetic Resolution

Chiral molecules with non-superimposable mirror images can have striking differences in biological activities, such as pharmacology, toxicology, pharmacokinetics, and metabolism [8]. In fact, more than half of the drugs currently in use are chiral compounds. In order to fulfill the increasing demand for enantiopure compounds, significant advances in asymmetric synthesis and catalysis have been achieved [9,10]. Dynamic kinetic resolution (DKR) catalyzed by transition-metal racemization complexes and kinetic resolution enzymes, have been employed as efficient methods to prepare chiral alcohols and amines that constitute important synthetic building blocks of various chemical products, such as agrochemicals, food additives, fragrances and pharmaceuticals [11]. Enzymatic kinetic resolution (KR) of racemic mixtures is the most common approach to access enantiomerically pure alcohols and amines on an industrial scale due to the high activity and selectivity of enzymes [12]. The resolution of racemic alcohols or amines is generally accomplished through (R)- or (S)- selective acylation of their enantiomers by using a lipase or a serine protease as the resolving enzyme. However, enzymatic KR suffers from the limitation that only 50% of the theoretical yield could be obtained for the desired enantiomer. Integrating a racemization catalyst to continuously replenish the consumed enantiomer could theoretically drive the resolution up to 100% (Scheme 1). The compatibility between the enzyme and the isomerization catalyst is essential for a successful DKR system. The KR enzyme must have sufficient enantioselectivity ($k_{fast}/k_{slow} \geq 20$) and the rate of isomerization ($k_{rac}$) must be at least 10 times faster than the enzyme-catalyzed reaction of the slow reacting enantiomer ($k_{slow}$).

Since the pioneering work of William [13], Bäckvall [14] and Kim [15] on developing practical systems that combined metal catalysts with lipases or serine proteases for DKR of alcohols and amines, a variety of studies have been performed to improve these systems, such as discovering catalysts that could efficiently racemize alcohols and amines at mild conditions, improving the stability and catalytic efficiency of enzymes, and expanding the substrate scopes [9,10]. To date, immobilized *Candida antarctica* lipase B (CALB) [16] and *C. antarctica* lipase A (CALA) [3] have been the most common enzymes of choice to prepare R- or S- enantiomers of alcohols respectively, owing to their robustness and activity in organic solvents at temperatures up to 100 °C.

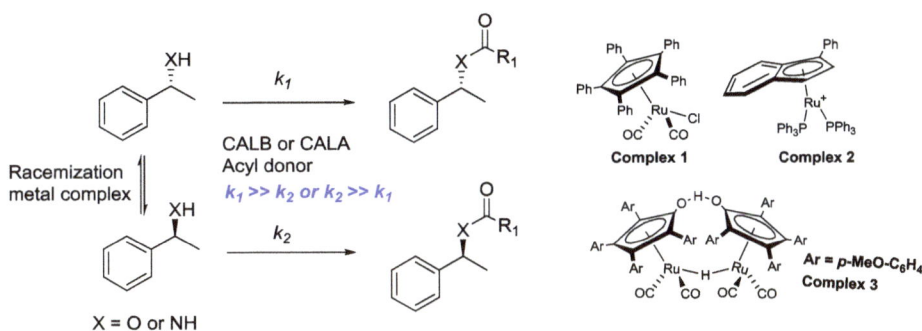

**Scheme 1.** An example of selective chemoenzymatic dynamic kinetic resolution (DKR) of secondary alcohols or primary amines by recently developed transition-metal complexes. CALB, *Candida antarctica* lipase B; CALA, *C. antarctica* lipase A.

## 2.1. Dynamic Kinetic Resolution of Secondary Alcohols and Derivatives

For the racemization of alcohols, the most commonly employed chemical catalysts are ruthenium complexes. Monomeric ruthenium pentamethylcyclopentadiene complex **1** developed by Bäckvall and co-workers [17] has been widely used in tandem with different enzymes to deracemize a wide range of functionalized secondary alcohols, including aliphatic alcohols [18,19], allylic alcohols [20–22], chlorohydins [23], diols [24,25], homoallylic alcohols [26], and *N*-heterocyclic 1,2-aminos alcohols [27] with excellent yields and enantiomeric excess (*ee*). Recently, complex **1** has been employed to synthesize biologically active 5,6-dihydropyran-2-ones and the corresponding δ-lactones [26]. Several new ruthenium complexes were developed to be active at room temperature for pairing with thermolabile enzymes. Nolan and co-workers recently reported cationic ruthenium indenyl complex **2** that could catalyze racemization of secondary alcohols without a strong base [28]. By coupling **2** with Novozyme® 435 (Strem, Boston, MA, USA) the DKR of various secondary alcohols was achieved in high yield and *ee* at room temperature. At the same time, the group of Martín-Matute found that a commercially available [Ru(*p*-cymene)Cl$_2$]$_2$ with the ligand 1,4-bis-(diphenylphosphino)butane could be coupled with a lipase from *Pseudomonas stutzeri* for the efficient DKR of α-hydroxyl ketones at ambient temperature [29]. The resulting enantiopure compounds provided straightforward access to a variety of diols and amino alcohols in a diastereo- and enatioselective manner.

Other more cost-effective and readily accessible metal complexes, such as iridium, aluminum and vanadium complexes, have also been investigated for DKR of secondary alcohols [9]. Akai and co-workers have demonstrated that the oxyvanadium (V) complex [VO(OSiPh$_3$)$_3$] in tandem with various lipases allowed for DKR of a wide range of linear and cyclic allylic secondary alcohols [30,31] (Scheme 2). Recently, they prepared a novel oxyvanadium catalyst (V-MPD), immobilized inside mesoporous silica (MPS) [32]. This heterogeneous catalyst could be recycled six times without any loss in activity, and it could mediate the racemization of benzylic, heteroaromatic and propargylic alcohols (Scheme 2).

The application of transition-metal catalysts in terms of DKR has been mainly limited to academic research due to their high cost and low stability. In fact, several heterogeneous and immobilized racemization catalysts have been made to improve the total turnover number (TON) of catalysts [10,32,33]. In addition, some heterogeneous acid catalysts, such as zeolite and nanozeolite microspheres, combined with immobilized lipases have been developed for efficient DKR of benzylic alcohols [10]. Very recently, Tang and co-workers reported a core-shell nanozeolite@enzyme bi-functional catalyst consisting of CALB immobilized on H-β zeolite microspheres coated with polydiallydimethylammonium chloride (PDDA) [34,35]. This core-shell structure modulated the optimum rate of racemization and KR to achieve the best catalytic performance. PDDA also protected the interaction between the products and the acidic core to minimize side products. However, this system was limited to kinetic resolution of benzylic alcohols.

**Selected examples of allyl alcohols deracemized by [VO(OSiPh₃)₃] and lipase**

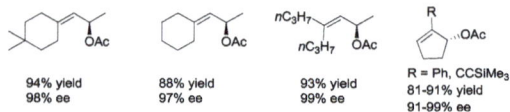

94% yield
98% ee

88% yield
97% ee

93% yield
99% ee

R = Ph, CCSiMe₃
81-91% yield
91-99% ee

**Selected examples of alcohols deracemized by immobolized V-MPD and lipase**

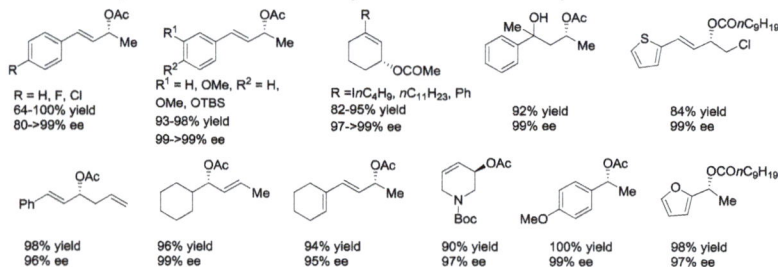

R = H, F, Cl
64-100% yield
80->99% ee

R¹ = H, OMe, R² = H,
OMe, OTBS
93-98% yield
99->99% ee

R =lnC₄H₉, nC₁₁H₂₃, Ph
82-95% yield
97->99% ee

92% yield
99% ee

84% yield
99% ee

98% yield
96% ee

96% yield
99% ee

94% yield
95% ee

90% yield
97% ee

100% yield
99% ee

98% yield
97% ee

**Scheme 2.** Scope of DKR systems involving vanadium complexes and various enzymes. V-MPD, novel oxyvanadium catalyst.

In addition to improving the catalytic efficiency of the racemization catalysts, a considerable amount of work has been carried out to improve the catalytic performance of lipases as biocatalysts through immobilization, cross-linking, surfactant stabilization or enzyme engineering. For example, Kim and co-workers recently coated lipoprotein lipase (LPL) from *Burkholderia* species with dextrin and ionic surfactant to produce activated lipoprotein lipase (LPL-D1) [36]. LPL-D1 was 3000-fold more active than its native protein in organic solvent and can facilitate DKR of diarylmethanols that had sub-optimal yields and enantiopurities. Bäckvall and co-workers applied a focused combinatorial gene mutagenesis technique to discover a mutant of *Candida antarctica* lipase A-Y93L/L367I with more than 30 times improvement on enantioselectivity of *sec*-alcohols in organic solvent [37]. More examples published before 2015 can be found in several reviews [9,10].

*2.2. Dynamic Kinetic Resolution of Amines*

The DKR of amines is more challenging due to a lack of efficient amine racemization catalysts. Amines are strong metal ligands; and high temperatures are utilized to prevent complexation of metals to the amines. In addition, highly active imine intermediates are likely to take part in several side reactions, which are more favored at elevated temperatures [9,10]. To date, a ruthenium complex Shvo analogue **3** coupled with CALB is the most practical method for DKR of aliphatic and benzylic primary amines at 90 °C [38–40].

In recent years, several palladium-based heterogeneous racemization catalysts have also been developed for DKR of benzylic primary amines. Bäckvall and co-workers developed a DKR system by using a catalyst consisting of palladium nanoparticles supported on amino-functionalized siliceous mesocellular foam (Pd(0)-Amp-MCF) to convert 1-phenylethylamine to an amide at 50 °C with sensitive Amano Lipase PS-C1 (*Burkholderia cepacia* lipase immobilized on ceramic beads) (Scheme 3). Pd-Amp-MCF is more efficient due to shorter reaction times and high TON compared with Pd nanoparticles alone [41]. Similarly, Li and co-workers investigated the effect of alkali salts on the catalytic activity of Pd nanoparticles on micro/mesoporous silica or activated carbon, and discovered that alkali salts greatly enhanced the selectivity of Pd catalysts [42]. Liang and coworkers used a similar system to synthesize rasagiline [43]. Moreover, a modified solvent extraction system and a continuous flow reactor were developed to compartmentalize the Pd-based nanoparticles and lipases to overcome the incompatibility of reaction conditions required for the racemization and enzymatic steps [44–46]. However, Pd nanocatalysts did not work well for aliphatic amines. Raney Ni and Co displayed preferences for aliphatic primary amine racemization (Scheme 3), but they have inhibitory effects on the enzyme, resulting in a slow DKR system [47].

**Selected examples of benzylic primary amines deracemized by Pd-Amp-MCF**

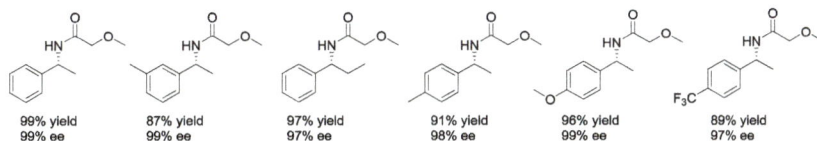

| | | | | | |
|---|---|---|---|---|---|
| 99% yield 99% ee | 87% yield 99% ee | 97% yield 97% ee | 91% yield 98% ee | 96% yield 99% ee | 89% yield 97% ee |

**Selected examples of aliphatic primary amines deracemized by Raney Ni and Co**

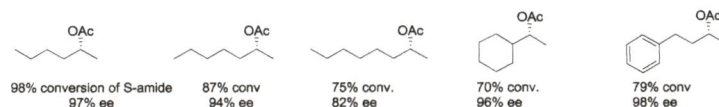

| | | | | |
|---|---|---|---|---|
| 98% conversion of S-amide 97% ee | 87% conv 94% ee | 75% conv. 82% ee | 70% conv. 96% ee | 79% conv 98% ee |

**Scheme 3.** Scope of DKR systems involving Pd(0)-Amp-MCF and Raney Ni or Co. Pd(0)-Amp-MCF, Pd(0)-aminopropyl-mesocellular foam.

## 2.3. Other Tandem Reactions by Transition Metal Catalysts and Lipases

Except for DKR of alcohols and amines, lipases have also been coupled with base catalysts or transition metal catalysts for synthesis of more complex compounds. Ramström and co-workers recently reported asymmetric synthesis of 1,3-oxanthiolan-5-one derivatives [48] and oxathiazinanones [49] through dynamic covalent kinetic resolution. In the first case, dynamic hemithioacetal formation combined with intramolecular, CALB-catalyzed lactonization resulted in the final product in good conversion with moderate to good enantiomeric excess (*ee*) (Scheme 4A). In the second case, CALB catalysis was coupled with a dynamic dominonitrone addition-cyclization pathway to synthesize new, six-membered N,O,S-containing heterocycles (Scheme 4B).

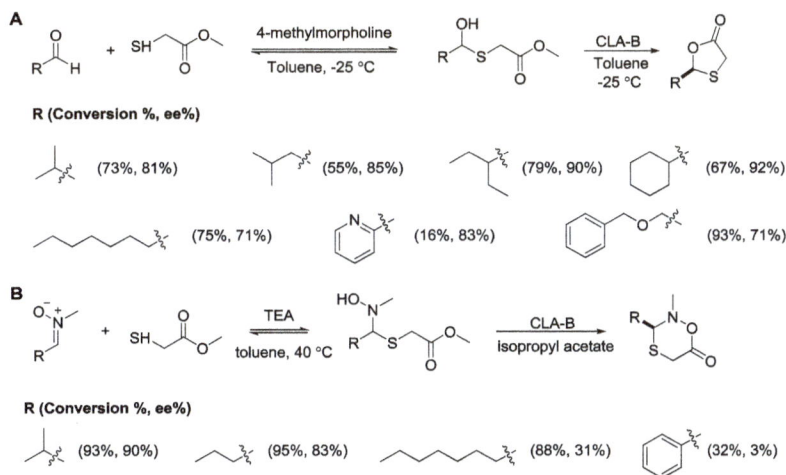

**Scheme 4.** Asymmetric synthesis of (**A**) 1,3-oxaathilan-5-one and (**B**) oxathiazinanones through dynamic covalent kinetic resolution. TEA, triethanolamine.

Most recently, Berglund and co-workers integrated heterogeneous Pd(0)-aminopropyl-mesocellular foam (Pd(0)-AmP-MCF) or Pd(0)-aminopropyl-conrolled pore glass (Pd(0)-AmP-CPG) and CALB catalysts in one-pot for eco-friendly and asymmetric synthesis of valuable molecules such as amines and amides from an aldehyde, ketone or an alcohol respectively in good to high overall yields [50]. In this work, they developed several novel cocatalytic relay sequences, including reductive amination/amidation (Scheme 5A), aerobic oxidation/reductive amination/amidation (Scheme 5B), and reductive amination/dynamic kinetic resolution (Scheme 5C).

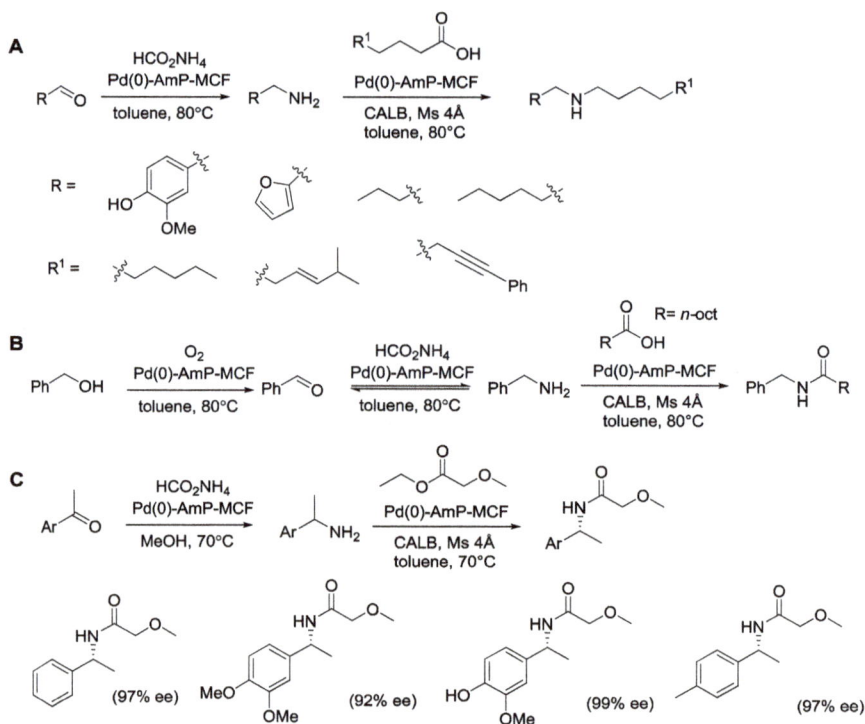

**Scheme 5.** Examples of integrated heterogeneous metal/enzymatic multiple relay catalysis for asymmetric synthesis: (**A**) reductive amination/amidation; (**B**) aerobic oxidation/reductive amination/amidation; (**C**) reductive amination/dynamic kinetic resolution.

## 3. One-Pot Chemoenzymatic Transformations

Except for the above-mentioned immobilized lipases and serine proteases, the majority of biocatalysts have poor stability in organic solvents and at high temperatures. To address the synthetic challenges of both chemistry and biology, there is an ongoing interest to combine chemocatalytic and biocatalytic reactions in a one-pot process in aqueous solution and at ambient temperatures. Use of water as a solvent is ecofriendly and reduces organic solvent usage and waste generation. It also provides accessibility to all enzymes in nature. Currently one-pot cascade or concurrent reactions avoid purification of intermediates and thus save time and cost. However, the development of such a system is generally hampered by incompatible reaction conditions, catalyst inhibition, undesired side reactions and poor solubility of substrates. Several approaches have been implemented to improve the one-pot operation of chemocatalysts and biocatalysts, including partitioning in a biphasic system, incorporation into a supramolecular cage, catalysts compartmentation, and development of artificial metalloenzymes [51,52].

## 3.1. Concurrent Tandem Reactions by Transition-Metal Complexes and Enzymes

Transition-metal catalysis and biocatalysis are two different disciplines in terms of synthesis. Some metal-catalyzed transformations cannot be accessed by enzymes, such as Pd-catalyzed cross coupling reactions [53], Wacker-oxidation [54], and Ru-catalyzed metathesis [55,56]. On the other hand, biocatalysis has strong selectivity on both substrates and products, enabling much cleaner reactions. The motivation to combine them in one-pot is to exploit the synthetic power that cannot be achieved by either of them separately.

There are a few successful examples of coupling Pd-catalyzed Suzuki cross-coupling reactions and Heck reactions in tandem with alcohol dehydrogenase in an aqueous buffer by using either a biphasic system [57] or additive tagged Pd-nanoparticles [58,59]. Recently, Zhao and coworkers successfully combined Ru-catalyzed olefin metathesis with a P450-monooxygenase catalyzed oxidation in a biphasic system for the synthesis of various epoxides [60]. Importantly, this work demonstrated the power of chemoenzymatic one-pot processes to achieve higher yields compared with the sequential two-step reactions. In this case, the cooperation of ruthenium metathesis and P450-monooxygenase led to a dynamic equilibrium of alkenes and a selective epoxidation of the cross-metathesis products **3** (Scheme 6A). The yield of **4** was 1.5 times higher than the hypothetical yield ($\leq$64%) resulting from stepwise reactions. Later, they developed a similar system by using an engineered P450 to convert a mixture of alkenes into a single aryl epoxide in high enantiomeric excess and moderate yield (Scheme 6B) [61].

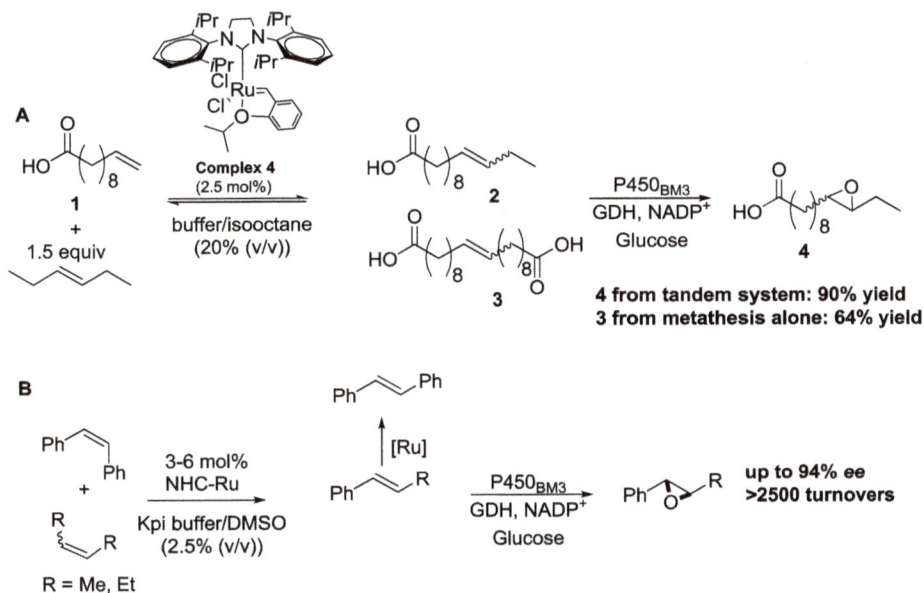

**Scheme 6.** Combination of an olefin metathesis with a P450 catalyzed epoxidation in a tandem-type one-pot process for the synthesis of (**A**) 10-undecenoic acid epoxide and (**B**) aryl epoxides. GDH, glucose dehydrogenase; NADP, nicotinamide adenine dinucleotide phosphate. Adapted from [61]. Copyright American Chemical Society, 2015.

The coordination of transition-metal complex to the enzyme is one of the major factors contributing to decreased catalytic abilities of both catalysts, thus hampering the progress of coupled tandem reactions [62,63]. Toste and co-workers addressed this limitation by encapsulating Au(I) or Ru(II) complexes in a $Ga_4L_6$ tetrahedral supramolecular cluster [62]. By doing so, they successfully coupled Au(I)-Ga$_4$L$_6$ host-guest with lipases and esterases for cascade hydrolysis, followed by a hydroalkoxylation of alkenes (Scheme 7). They also achieved a Ru(II)-mediated olefin isomerization of 2-propen-1-ol to give propanal, followed by reduction to propanol by an alcohol dehydrogenase (ADH). In both cases, the yields were improved relative to applying free cationic catalysts directly.

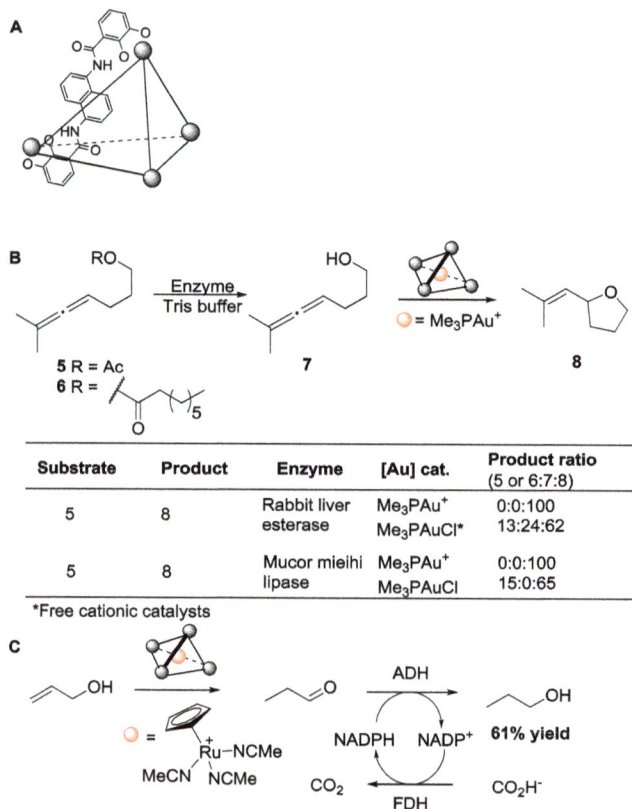

| Substrate | Product | Enzyme | [Au] cat. | Product ratio (5 or 6:7:8) |
|---|---|---|---|---|
| 5 | 8 | Rabbit liver esterase | Me$_3$PAu$^+$ | 0:0:100 |
|   |   |   | Me$_3$PAuCl* | 13:24:62 |
| 5 | 8 | Mucor mieihi lipase | Me$_3$PAu$^+$ | 0:0:100 |
|   |   |   | Me$_3$PAuCl | 15:0:65 |

*Free cationic catalysts

**Scheme 7.** A schematic view of the Ga$_4$L$_6$ tetrahedral supramolecular assembly (**A**) in which each edge of the tetrahedron represents a bisbidentate ligand and each vertex represents a gallium center (grey balls); (**B,C**) are metal catalysts encapsulated within a supramolecular complex in tandem with biocatalysts. ADH, alcohol dehydrogenase; FDH, formaldehyde dehydrogenase. Reprinted by permission from Macmillan Publishers Ltd.: Nature Chemistry [62]. Copyright 2013.

Catalysts compartmentation is another useful strategy to solve several incompatibility problems when combining heterogeneous organic or inorganic catalysis with enzyme catalysis. Careful reaction design led to efficient chemoenzymatic transformations catalyzed by immobilized chemical catalysts and biocatalysts, which are utilized in different compartments. Those multi-pot reactions were generally ecofriendly by avoiding intermediate isolation steps [64–66]. Recently, one-pot compartmentations also have been achieved by using membrane or encapsulation techniques. Gröger and co-workers combined noncompatible CuCl/PdCl$_2$-catalyzed Wacker oxidation with alcohol dehydrogenase-catalyzed ketone reduction to convert styrene enantioselectively into 1-phenyl-ethanol in one-pot with good conversion and *ee* (Scheme 8) [67]. To overcome mutual inactivation of both catalysts, Wacker oxidation was conducted in the interior of a polydimethylsiloxane (PDMS) thimble that enabled the diffusion of only the organic substrate and product into the exterior where the biotransformation takes place (Scheme 8). In another example, they developed a one-pot cascade reaction combining a co-factor free decarboxylase from *Bacillus subtilis* named *bs*PAD with a Ru metathesis catalyst to produce high-value antioxidants in good yield from bio-based precursors [68]. Encapsulation of *bs*PAD in an aqueous environment created by poly(vinyl alcohol)/poly(ethylene glycol) (PAV/PEG) cryogels enabled the enzyme functionalize in pure organic solvent (Scheme 9). Compartmentation not only overcame the catalytic incompatibility issue, but also realized the recycling of valuable catalysts in a more convenient manner.

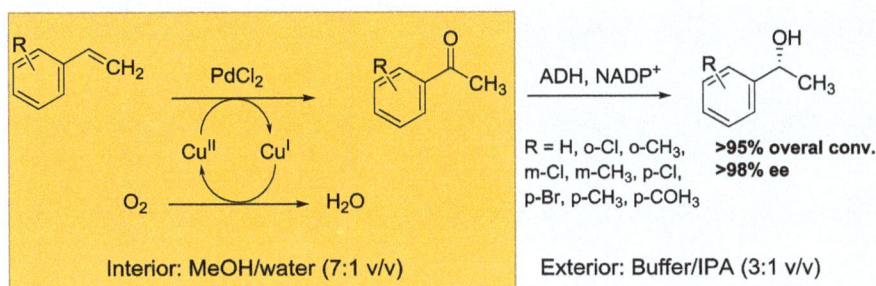

**Scheme 8.** Combination of Wacker oxidation and enzymatic reduction in one-pot aqueous media through compartmentalization. Adapted from [67]. Copyright WILEY-VCH Verlag GmbH & Co. KGaA, Weinheim, 2015.

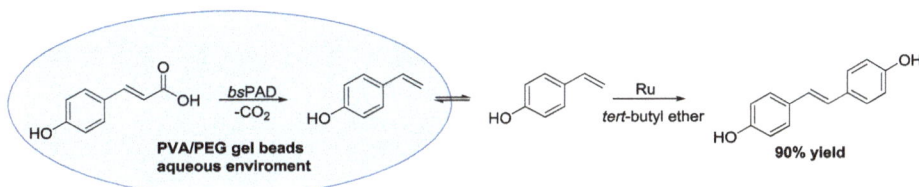

**Scheme 9.** One-pot cascade reaction combining an encapsulated decarboxylase with a metathesis catalyst for the synthesis of bio-based antioxidants. PVA, poly(vinyl alcohol); PEG, poly(ethylene glycol); *bs*PAD, co-factor free decarboxylase from *Bacillus subtilis*.

### 3.2. Coupling Visible-Light Photoredox Catalysis with Biocatalysts

Visible-light photoredox catalysis has been considered a highly desirable process in response to the interest in renewable energy and green chemistry. Organic transformations are afforded by combining photoredox with transition-metal complexes, or electrocatalysis [69]. Recently, photoredox has entered the realm of biocatalysis to form a more synergistic framework for catalysis of challenging reactions under mild conditions. In these examples, biotransformations of organic molecules are driven by the energy from the photoredox-catalyzed, light-dependent process.

Cheruzel and co-workers developed hybrid P450 BM3 heme domains containing a covalently attached Ru(II) photosensitizer to afford light-driven hydroxylation of lauric acid with improved TON and initial reaction rate compared with normal P450 systems (Scheme 10) [70]. This process also circumvented the use of reductases and NAD(P)H cofactors by employing Ru(II) mediated electron transfer processes to reduce Fe(III) to Fe(II) in the presence of dithiocarbamate (DTC). In a follow-up study, Park and co-workers created a cofactor-free light-driven whole cell cytochrome P450 catalysis for the bioconversion of various substrates, including marketed drugs simvastatin® (Merck & Co., Inc., Kenilworth, NJ, USA), lovastatin and omeprazole [71]. Instead of transition-metal photoredox, an organic dye, eosin Y that can easily enter the cytoplasm of *Escherichia coli* (*E. coli*) and specifically bind to the heme domain of P450 was used. Under visible-light irradiation, the reductive quenching of excited eosin Y resulted in electron transfer to Fe(III) in the P450 heme domain. The activated P450 conducted selective organic transformations that were controlled by the catalytic cycle of P450. This work demonstrated a whole cell platform for co-factor free, reductase-independent P450 photocatalysis.

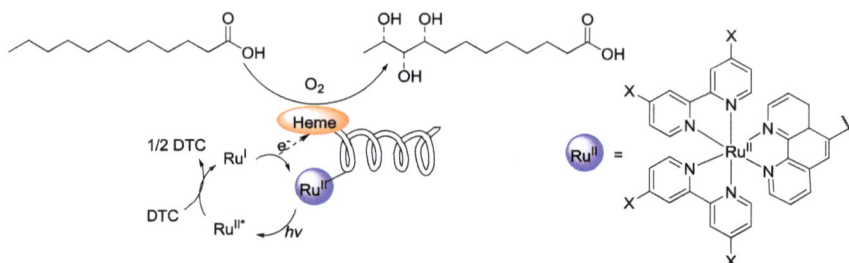

**Scheme 10.** Cooperative Ru(II) (X = H or OMe) photoredox catalysis and P450 BM3 biocatalysis. Adapted from [70]. Copyright American Chemical Society, 2013.

Instead of coupling with oxidation enzymes, photoredox catalysts, working as hydride transfer catalysts, have also been incorporated with reductases such as alcohol dehydrogenase [72] and glucose dehydrogenase [73] to prepare chiral alcohols and L-glutamate respectively. Unlike the aforementioned studies in which organic sacrificial electron donors were used, Corma and co-workers used light-driven and titanium dioxide-promoted water oxidation to drive redox reactions catalyzed by flavin-based old yellow enzyme (OYE) (Scheme 11) [74]. The protons and electrons were generated by Au/TiO$_2$ photoredox catalyst via the oxidation of water under UV irradiation, and then were supplied to the flavin of old yellow enzymes for asymmetrically reducing conjugated C=C bonds. As the oxidation of water to oxygen is the rate-limiting step of the current process, there should be room for improvement with some optimization. Most recently, Ru(II) and Ir(II) complexes were successfully applied as photosensitizers for regeneration of nicotinamide adenine dinucleotide phosphate (NADPH) in OYE catalytic cycles. However, those systems required an extra sacrificial electron donor, triethanolamine [75].

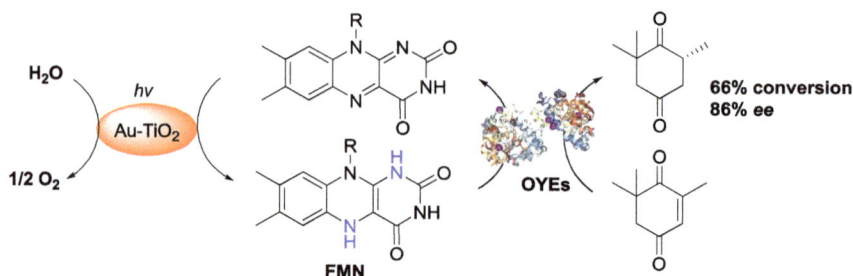

**Scheme 11.** TiO$_2$ and OYE-based photoenzymatic reduction of ketoisophorone. FMN, flavin mononucleotide, OYE, old yellow enzyme. Reprinted with permission from Macmillan Publishers Ltd.: Nature Chemistry [74]. Copyright 2014.

## 3.3. Artificial Metalloenzymes for Selective Transformations

An intense area of research has been the generation of artificial metalloenzymes by incorporating metal catalysts into protein scaffolds such as streptavidin [76], bovine serum albumin, and apo-myoglobin [77]. With the aid of computational, molecular and structural biology, functional artificial metalloenzymes have been developed by de novo design or protein redesign processes [78]. Not limited to mimicking naturally occurring transformations, biochemists are currently trying to generate artificial metalloenzymes with catalytic abilities that have no equivalents in either chemical catalysis or biocatalysis. Additionally, similar to the incorporation of transition-metal complexes within supramolecular host-guest complexes, the use of artificial metalloenzymes enables the catalytic activity of transition-metal catalysts in biological environments, allowing synthetic catalysts to work collaboratively with other enzymes. There are several recent reviews discussing the newest techniques and examples of various artificial metalloenzymes [51,76,78–84]. To avoid duplication, we will only highlight the major accomplishments since 2015 in terms of tandem catalysis.

One of the biggest motivations to develop artificial metalloenzymes is to access both the reactivity and substrate scope of metal complexes together with the regio-, stereo- and enantio-selectivities afforded by the protein scaffolds. Hartwig and co-workers reported a concise method of replacing iron in Fe-porphyrin IX (Fe-PIX) proteins with abiological, noble metals to create enzymes that can catalyze reactions not catalyzed by native Fe-enzymes or other metalloenzymes [85]. They conducted directed evolution of a modified myoglobin containing an Ir(Me) site. Impressively, the resulting mutants could catalyze enantioselective C–C bond formation through carbene insertion and also the enantio-diastereoselective cyclopropanation of unactivated olefins (Scheme 12). This method sets the stage for the generation of artificial enzymes from innumerable combinations of PIX—proteins scaffolds and unnatural metal cofactors for various abiological transformations.

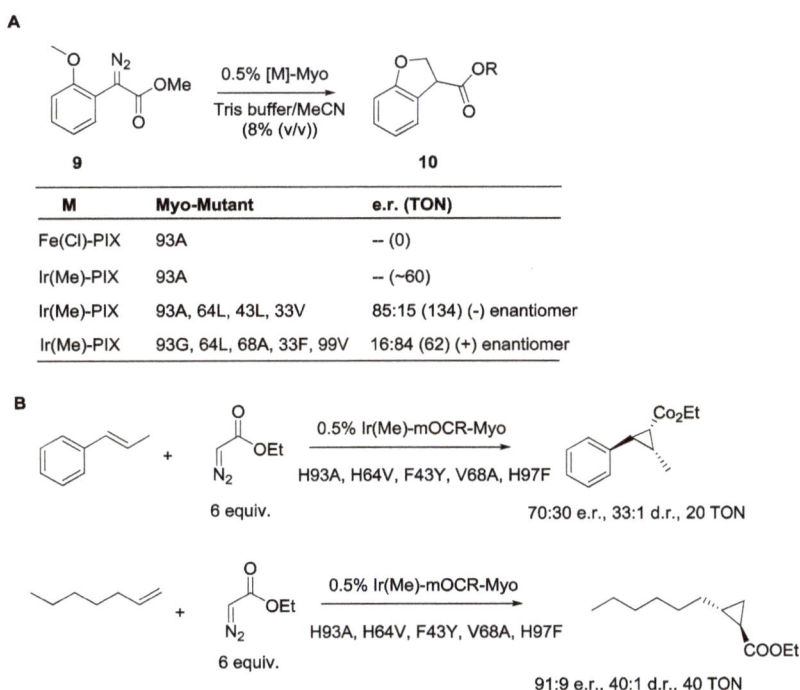

**Scheme 12.** Artificial Ir(Me)-based myoglobin-catalyzed C–H insertion for C–C bond formation and carbene addition to internal or aliphatic olefins: (**A**) insertion carbenes into C–H bonds; (**B**) carbene addition to internal or aliphatic olefins. Reprinted with permission from Macmillan Publishers Ltd.: Nature [85]. Copyright 2016.

Instead of performing directed evolution in vitro, Ward and coworkers utilized compartmentalization and in vivo evolution of a streptavidin (SAV)-biotin based artificial metalloenzyme for olefin metathesis (Figure 1) [86]. They created an *E. coli* strain for periplasmic expression of SAV with a biotinylated Hoveyda-Grubbs catalyst. The periplasm offered an auspicious environment for artificial metalloenzymes and facilitated artificial metathase-catalyzed metathesis in vivo and its directed evolution in a high throughput manner. This strategy not only created an artificial metalloenzyme comparable with commercial catalysts, showing activity for different metathesis substrates, but also represented the systematic implementation and evolution of an artificial metalloenzyme that catalyzes an abiotic reaction in vivo with other potential applications, such as non-natural metabolism.

Similar to the supramolecular complex, the protein scaffolds compartmentalize the chemical catalysts and avoid mutual inactivation of transition metal complexes and enzymes. By embedding a biotinylated $d^6$-Ir pianostood complex within SAV, Ward and co-workers enabled Ir-based transfer hydrogenation in the presence of *E. coli* cell free extracts and cell lysates [87]. In addition, they applied a similar strategy to create an artificial transfer hydrogenase (ATHase) that was successfully coupled with various NADH-, FAD- and heme-dependent enzymes for orthogonal redox cascade reactions that could not have been generated when free Ir-complex was used [88]. Significantly, by coupling ATHase with monoamine oxidase (MAO-N), NADPH regeneration was achieved and L-pipecolic acid was prepared with 99% *ee* (Scheme 13).

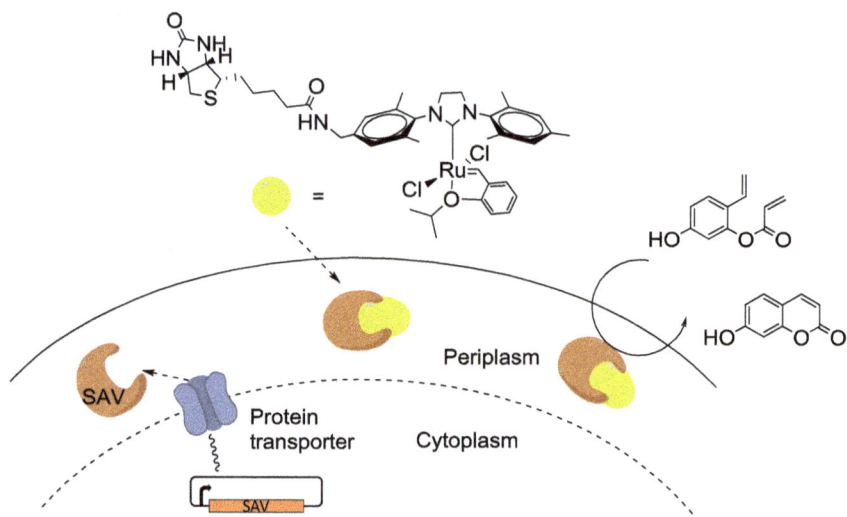

**Figure 1.** Streptavidin (SAV)-based artificial metalloenzymes for in vivo metathesis. The dashed arrows indicate the transportation of the chemical catalyst and streptavidin into the periplasm. Reprinted with permission from Macmillan Publishers Ltd.: Nature [86]. Copyright 2016.

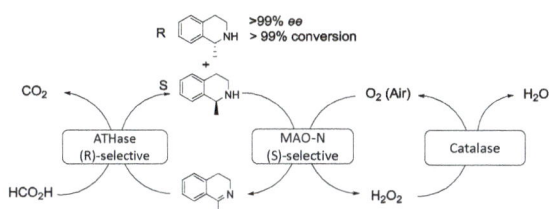

Selected examples of orthogonal redox cascades combining ATHase with oxidase or oxygenase

| Substrate | Prodcut | Conv. (%) | ee (%) |
|---|---|---|---|
| | | >99% | >99% |
| | | 98% | 99% |
| | | 98% | 99% |
| | | 99% | >99% |
| | | 65% | 99% |
| | | 88% | 86% |

**Scheme 13.** Enzyme cascade for the double stereoselective deracemization of amines. Reprinted with permission from Macmillan Publishers Ltd.: Nature Chemistry [88]. Copyright 2013.

## 4. Interfacing the Transition-Metal Catalysis with Living Cells

The use of whole cells as a catalytic "factory" to synthesize fine chemicals, pharmaceuticals, and steroids is an emerging area due to several advantages. The cell could provide natural protection for the proteins and thus improve their catalytic turnover. The compartmentalization also enables concurrent chemoenzymatic reactions to occur without inactivation due to incompatibility of the biocatalysts and other transition-metal complexes [3,89,90]. There has been much progress in engineering multi-enzymatic steps in cells for chemical production by applying metabolic engineering strategies [91]. However, the adaption of metal catalysis to the whole cell catalytic system remains widely unstudied mainly due to the deactivation of metal complexes under biological conditions. We will highlight recent examples in this area and the progress towards bio-orthogonal catalysis with organometallic compounds.

Other than the example discussed in Section 3.2 about cofactor-free light-driven whole cell cytochrome P450 catalysis, a biometallic whole cell catalyst for enantioselective deracemization of secondary amines was engineered by Lloyd and co-workers [92]. The engineered aerobic cultures of *E. coli*, overproducing a recombinant monoamine oxidase (MAO-N-D5) possessing high enantioselectivity against chiral amines, were coated with nanoscale Pd(0) precipitated via bioreduction reactions (Scheme 14). The whole cell catalyst was prepared (*R*)-1-methyltetrahydroisoquinoline (MTQ) with 96% *ee* by a dynamic process, in which MAO selectively oxidized *S* enantiomer of racemic mixture. The resulting 1-methyl-3,4-dihydroisoquinoline (MDQ) was then reduced back to the racemic amine by nonselective Pd/$H_2$ reduction. This work is important for the preparation of chiral secondary amines that are hard to obtain via normal metal-lipase DKR system. More recently, Balskus and co-workers reported a method for alkene hydrogenation that utilized the Royer Pd catalyst [93] and hydrogen gas generated directly by an engineered *E. coli* (Scheme 15) [94]. This work first demonstrated that the metabolic output of living microbes and a biocompatible non-enzymatic transformation could be combined to enable preparative scale chemical synthesis.

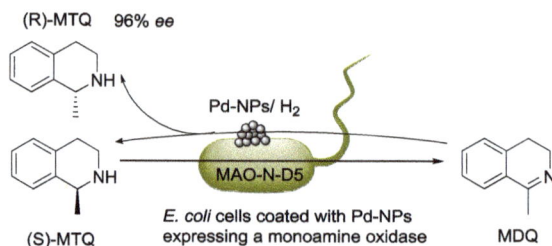

**Scheme 14.** Deracemization of a cyclic secondary amine by engineered biometallic whole cell catalyst. MTQ, methyltetrahydroisoquinoline; MDQ, 1-methyl-3,4-dihydroisoquinoline; MAO-N-D5, monoamine oxidase. Adapted from [92]. Copyright American Chemical Society, 2011.

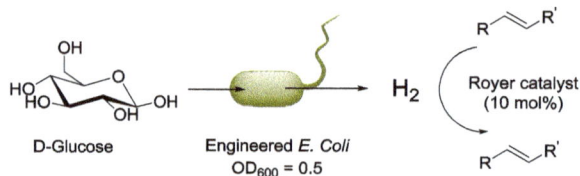

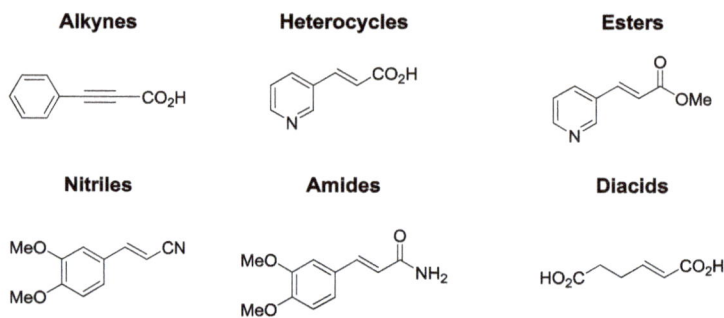

**Scheme 15.** A biocompatible alkene hydrogenation combines organic synthesis with microbial metabolism. Adapted from [94]. Copyright WILEY-VCH Verlag GmbH & Co. KGaA, Weinheim, 2014.

Considerable attention has been focused on improving the biocompatibility of transition-metal complex-catalyzed bio-orthogonal reactions in living cells. Currently, bio-orthogonal reactions have many applications for selective labeling and modification of biomolecules in living systems [95]. Except for the work by Ward and coworkers mentioned in Section 3.3, there are few examples of using such techniques for chemical production. However those results made researchers more aware of the need to prepare more biocompatible transition-metal complexes that are able to function in living cells. For instance, Eric and co-workers published a set of organometallic Ru complexes for the catalytic uncaging of allylcarbamate (alloc)-protected amines within mammalian cells. They applied this method in activating a caged anticancer drug, which efficiently induced apoptosis in HeLa cells [96]. Recently Mascareñas and co-workers improved this scenario further by using a designed Ru complex that accumulated preferentially inside the mitochondria of mammalian cells while keeping its ability to uncage alloc-protected amines [97]. Except for the Ru catalysts, Pd complexes have also been applied in living cells for activation of intracellularly lysine-based proteins by decaging a propargyloxycarbonyl (Proc)-caged lysine analogue [98].

## 5. Future Prospects and Conclusions

The combination of chemocatalysis and biocatalysis in a one-pot for concurrent transformations is still in its infancy. This is particularly true for chemoenzymatic transformations in aqueous media. DKR catalyzed by transition-metal complexes and lipases or serine proteases have been fully developed to efficiently prepare most enantiopure primary and secondary alcohols. However, more advanced racemization catalysts should be developed for more challenging substrates such as chlorohydrins, alcohols with distant olefin groups and tertiary alcohols. In contrast to alcohols, there are significantly fewer available DKR systems for amines, especially aliphatic amines and secondary amines, due to the lack of an efficient racemization strategy. The examples of one-pot tandem chemoenzymatic reactions involving other enzymes in aqueous media are even sparser, and development of this field has been slow, mainly due to the mutual inactivation of metal complexes and enzymes. Unlike multistep enzymatic transformations occurring in cells, resulting in complex molecules, the tandem processes developed so far generally consist of two or three catalytic steps for simple molecule synthesis. To address those issues, it is essential to take advantage of several research areas including, but not limited to, protein engineering, chemical catalyst synthesis, supramolecular assembly, artificial metalloenzymes and whole cell catalysis.

Synthetic chemists have been focusing on developing water-soluble transition-metal catalysts that are effective in aqueous solutions. The progress in this area has enabled several metal-catalyzed reactions in buffer solutions (e.g., olefin metathesis [99–101], Pd-catalyzed hydrogenation [102], and C–C coupling reactions [103]). Several methods have also been applied for improving enzymes as biocatalysts. Immobilization of enzymes on heterogeneous catalysts [104–106] and preparation of cross-linked enzyme aggregates [107] are well-developed methods to improve the activity and stability of enzymes at extreme conditions, and to facilitate enzyme recycling. In addition, engineering enzymes by rational design [108,109] and directed evolution [110] are alternative ways to obtain enzyme mutants with improved properties and new functions. The combination of these techniques will enable integration of chemocatalysis and biocatalysis for advanced synthesis.

Inspired by naturally occurring reactions processed in different organelles, compartmentalization is becoming an attractive strategy to avoid mutual inactivation between catalysts from different disciplines. Creation of supramolecular hosts with hydrophobic cavities and a transition metal complex is a very appealing strategy for controlling the metal complex properties in competitive water solvents containing proteins and other cellular components [111]. Similarly, artificial metalloenzymes have been further developed to integrate transition-metal catalysts into cascade reactions with other biocatalysts and foster bioorthogonal transformations catalyzed by metal complexes in living cells. Importantly, generating artificial metalloenzymes is an advanced way to create catalysts with novel activities to address more challenging transformations. Protein engineering strategies can be further applied to artificial metalloenzyme development to improve its catalytic performance.

Finally, whole cell fermentation is a well-known technique to produce metabolite-related chemicals. However, there are very few known processes that integrate bio-orthogonal reactions catalyzed by transition-metal catalysts into multi-step biocatalytic systems in cells for complex transformations. The progress of developing biocompatible transition-metal complexes that function in cells, together with strategies of engineering enzymatic multi-step catalysis in vivo [112,113], will open new windows for creating new cell factories for chemical production.

In conclusion, interest in combining biocatalysts and chemical catalysts continues to grow. With recent advances in protein engineering, catalyst synthesis, artificial metalloenzymes and supramolecular assembly, there is a great potential to develop sophisticated tandem chemoenzymatic processes for synthesis of complex chemicals in an ecofriendly manner. Therefore, more accomplishments in this area are expected in the near future.

**Acknowledgments:** This work was supported by U.S. National Science Foundation under the CCI Center for Enabling New Technologies through Catalysis (CENTC) Phase II Renewal, CHE-1205189. Yajie Wang is grateful for a graduate research fellowship from 3M.

**Author Contributions:** Yajie Wang and Huimin Zhao wrote the paper.

**Conflicts of Interest:** The authors declare no conflict of interest.

# References

1. Sheldon, R.A. *Multi-Step Enzyme Catalysis: Biotransformations and Chemoenzymatic Synthesis*; Wiley-VCH Verlag GmbH & Co. KGaA: Weinheim, Germany, 2008; p. 256.
2. Philip, J.P.; Clive, S.P.; Adrian, J.S. Tandem reactions in organic synthesis: Novel strategies for natural product elaboration and the development of new synthetic methodology. *Chem. Rev.* **1996**, *96*, 195–206.
3. Denard, C.A.; Hartwig, J.F.; Zhao, H. Multistep one-pot reactions combining biocatalysts and chemical catalysts for asymmetric synthesis. *ACS Catal.* **2013**, *3*, 2856–2864. [CrossRef]
4. Wells, A.S.; Finch, G.L.; Michels, P.C.; Wong, J.W. Use of enzymes in the manufacture of active pharmaceutical ingredients—A science and safety-based approach to ensure patient safety and drug quality. *Org. Process Res. Dev.* **2012**, *16*, 1986–1993. [CrossRef]
5. Winkler, C.K.; Tasnádi, G.; Clay, D.; Hall, M.; Faber, K. Asymmetric bioreduction of activated alkenes to industrially relevant optically active compounds. *J. Biotechnol.* **2012**, *162*, 381–389. [CrossRef] [PubMed]
6. Winkler, C.K.; Clay, D.; Davies, S.; O'Neill, P.; McDaid, P.; Debarge, S.; Steflik, J.; Karmilowicz, M.; Wong, J.W.; Faber, K. Chemoenzymatic asymmetric synthesis of pregabalin precursors via asymmetric bioreduction of β-cyanoacrylate esters using ene-reductases. *J. Org. Chem.* **2013**, *78*, 1525–1533. [CrossRef] [PubMed]
7. Savile, C.K.; Janey, J.M.; Mundorff, E.C.; Moore, J.C.; Tam, S.; Jarvis, W.R.; Colbeck, J.C.; Krebber, A.; Fleitz, F.J.; Brands, J.; et al. Biocatalytic asymmetric synthesis of chiral amines from ketones applied to sitagliptin manufacture. *Science* **2010**, *329*, 305–309. [CrossRef] [PubMed]
8. Nguyen, L.A.; He, H.; Pham-Huy, C. Chiral drugs: An overview. *Int. J. Biomed. Sci.* **2006**, *2*, 85–100. [PubMed]
9. Verho, O.; Backvall, J.E. Chemoenzymatic dynamic kinetic resolution: A powerful tool for the preparation of enantiomerically pure alcohols and amines. *J. Am. Chem. Soc.* **2015**, *137*, 3996–4009. [CrossRef] [PubMed]
10. De Miranda, A.S.; Miranda, L.S.; de Souza, R.O. Lipases: Valuable catalysts for dynamic kinetic resolutions. *Biotechnol. Adv.* **2015**, *33*, 372–393. [CrossRef] [PubMed]
11. Bartoszewicz, A.; Ahlsten, N.; Martin-Matute, B. Enantioselective synthesis of alcohols and amines by iridium-catalyzed hydrogenation, transfer hydrogenation, and related processes. *Chemistry* **2013**, *19*, 7274–7302. [CrossRef] [PubMed]
12. Breuer, M.; Ditrich, K.; Habicher, T.; Hauer, B.; Kesseler, M.; Sturmer, R.; Zelinski, T. Industrial methods for the production of optically active intermediates. *Angew. Chem. Int. Ed.* **2004**, *43*, 788–824. [CrossRef] [PubMed]
13. Allen, J.V.; Williams, J.M.J. Dynamic kinetic resolution with enzyme and palladium combinations. *Tetrahedron Lett.* **1996**, *37*, 1859–1862. [CrossRef]
14. Pamies, O.; Backvall, J.E. Combination of enzymes and metal catalysts. A powerful approach in asymmetric catalysis. *Chem. Rev.* **2003**, *103*, 3247–3262. [CrossRef] [PubMed]
15. Kim, M.J.; Chung, Y.I.; Choi, Y.K.; Lee, H.K.; Kim, D.; Park, J. (S)-Selective dynamic kinetic resolution of secondary alcohols by the combination of subtilisin and an aminocyclopentadienylruthenium complex as the catalysts. *J. Am. Chem. Soc.* **2003**, *125*, 11494–11495. [CrossRef] [PubMed]
16. Martin-Matute, B.; Backvall, J.E. Dynamic kinetic resolution catalyzed by enzymes and metals. *Curr. Opin. Chem. Biol.* **2007**, *11*, 226–232. [CrossRef] [PubMed]
17. Warner, M.C.; Backvall, J.E. Mechanistic aspects on cyclopentadienylruthenium complexes in catalytic racemization of alcohols. *Acc. Chem. Res.* **2013**, *46*, 2545–2555. [CrossRef] [PubMed]
18. Martin-Matute, B.; Edin, M.; Bogar, K.; Backvall, J.E. Highly compatible metal and enzyme catalysts for efficient dynamic kinetic resolution of alcohols at ambient temperature. *Angew. Chem. Int. Ed.* **2004**, *43*, 6535–6539. [CrossRef] [PubMed]

19. Martin-Matute, B.; Edin, M.; Bogar, K.; Kaynak, F.B.; Backvall, J.E. Combined ruthenium(II) and lipase catalysis for efficient dynamic kinetic resolution of secondary alcohols. Insight into the racemization mechanism. *J. Am. Chem. Soc.* **2005**, *127*, 8817–8825. [CrossRef] [PubMed]
20. Lihammar, R.; Millet, R.; Backvall, J.E. Enzyme- and ruthenium-catalyzed dynamic kinetic resolution of functionalized cyclic allylic alcohols. *J. Org. Chem.* **2013**, *78*, 12114–12120. [CrossRef] [PubMed]
21. Norinder, J.; Bogar, K.; Kanupp, L.; Backvall, J.E. An enantioselective route to alpha-methyl carboxylic acids via metal and enzyme catalysis. *Org. Lett.* **2007**, *9*, 5095–5098. [CrossRef] [PubMed]
22. Bogar, K.; Vidal, P.H.; Leon, A.R.; Backvall, J.E. Chemoenzymatic dynamic kinetic resolution of allylic alcohols: A highly enantioselective route to acyloin acetates. *Org. Lett.* **2007**, *9*, 3401–3404. [CrossRef] [PubMed]
23. Traff, A.; Bogar, K.; Warner, M.; Backvall, J.E. Highly efficient route for enantioselective preparation of chlorohydrins via dynamic kinetic resolution. *Org. Lett.* **2008**, *10*, 4807–4810. [CrossRef] [PubMed]
24. Leijondahl, K.; Boren, L.; Braun, R.; Backvall, J.E. Enzyme- and ruthenium-catalyzed dynamic kinetic asymmetric transformation of 1,5-diols. Application to the synthesis of (+)-Solenopsin A. *J. Org. Chem.* **2009**, *74*, 1988–1993. [CrossRef] [PubMed]
25. Leijondahl, K.; Boren, L.; Braun, R.; Backvall, J.E. Enantiopure 1,5-diols from dynamic kinetic asymmetric transformation. Useful synthetic intermediates for the preparation of chiral heterocycles. *Org. Lett.* **2008**, *10*, 2027–2030. [CrossRef] [PubMed]
26. Warner, M.C.; Shevchenko, G.A.; Jouda, S.; Bogar, K.; Backvall, J.E. Dynamic kinetic resolution of homoallylic alcohols: Application to the synthesis of enantiomerically pure 5,6-dihydropyran-2-ones and delta-lactones. *Chemistry* **2013**, *19*, 13859–13864. [CrossRef] [PubMed]
27. Traff, A.; Lihammar, R.; Backvall, J.E. A chemoenzymatic dynamic kinetic resolution approach to enantiomerically pure (*R*)- and (*S*)-duloxetine. *J. Org. Chem.* **2011**, *76*, 3917–3921. [CrossRef] [PubMed]
28. Fernandez-Salas, J.A.; Manzini, S.; Nolan, S.P. A cationic ruthenium complex for the dynamic kinetic resolution of secondary alcohols. *Chemistry* **2014**, *20*, 13132–13135. [CrossRef] [PubMed]
29. Agrawal, S.; Martinez-Castro, E.; Marcos, R.; Martin-Matute, B. Readily available ruthenium complex for efficient dynamic kinetic resolution of aromatic alpha-hydroxy ketones. *Org. Lett.* **2014**, *16*, 2256–2259. [CrossRef] [PubMed]
30. Akai, S.; Tanimoto, K.; Kanao, Y.; Egi, M.; Yamamoto, T.; Kita, Y. A dynamic kinetic resolution of allyl alcohols by the combined use of lipases and [VO(OSiPh$_3$)$_3$]. *Angew. Chem. Int. Ed.* **2006**, *45*, 2592–2595. [CrossRef] [PubMed]
31. Akai, S.; Hanada, R.; Fujiwara, N.; Kita, Y.; Egi, M. One-pot synthesis of optically active allyl esters via lipase-vanadium combo catalysis. *Org. Lett.* **2010**, *12*, 4900–4903. [CrossRef] [PubMed]
32. Egi, M.; Sugiyama, K.; Saneto, M.; Hanada, R.; Kato, K.; Akai, S. A mesoporous-silica-immobilized oxovanadium cocatalyst for the lipase-catalyzed dynamic kinetic resolution of racemic alcohols. *Angew. Chem. Int. Ed.* **2013**, *52*, 3654–3658. [CrossRef] [PubMed]
33. Nieguth, R.; ten Dam, J.; Petrenz, A.; Ramanathan, A.; Hanefeld, U.; Ansorge-Schumacher, M.B. Combined heterogeneous bio- and chemo-catalysis for dynamic kinetic resolution of (*rac*)-benzoin. *RSC Adv.* **2014**, *4*, 45495–45503. [CrossRef]
34. Li, X.; Yan, Y.; Wang, W.; Zhang, Y.; Tang, Y. Activity modulation of core and shell in nanozeolite@enzyme bi-functional catalyst for dynamic kinetic resolution. *J. Colloid Interface Sci.* **2015**, *438*, 22–28. [CrossRef] [PubMed]
35. Wang, W.; Li, X.; Wang, Z.; Tang, Y.; Zhang, Y. Enhancement of (stereo)selectivity in dynamic kinetic resolution using a core-shell nanozeolite@enzyme as a bi-functional catalyst. *Chem. Commun.* **2014**, *50*, 9501–9504. [CrossRef] [PubMed]
36. Lee, J.; Oh, Y.; Choi, Y.K.; Choi, E.; Kim, K.; Park, J.; Kim, M.-J. Dynamic kinetic resolution of diarylmethanols with an activated lipoprotein lipase. *ACS Catal.* **2015**, *5*, 683–689. [CrossRef]
37. Wikmark, Y.; Svedendahl Humble, M.; Backvall, J.E. Combinatorial library based engineering of *candida antarctica* lipase a for enantioselective transacylation of *sec*-alcohols in organic solvent. *Angew. Chem. Int. Ed.* **2015**, *54*, 4284–4288. [CrossRef] [PubMed]
38. Goossen, L.J.; Paetzold, J. Pd-catalyzed decarbonylative olefination of aryl esters: Towards a waste-free heck reaction. *Angew. Chem. Int. Ed.* **2002**, *41*, 1237–1241. [CrossRef]

39. Thalen, L.K.; Zhao, D.; Sortais, J.B.; Paetzold, J.; Hoben, C.; Backvall, J.E. A chemoenzymatic approach to enantiomerically pure amines using dynamic kinetic resolution: Application to the synthesis of norsertraline. *Chemistry* **2009**, *15*, 3403–3410. [CrossRef] [PubMed]
40. Thalen, L.K.; Backvall, J.E. Development of dynamic kinetic resolution on large scale for (+/−)-1-phenylethylamine. *Beilstein J. Org. Chem.* **2010**, *6*, 823–829. [CrossRef] [PubMed]
41. Gustafson, K.P.; Lihammar, R.; Verho, O.; Engstrom, K.; Backvall, J.E. Chemoenzymatic dynamic kinetic resolution of primary amines using a recyclable palladium nanoparticle catalyst together with lipases. *J. Org. Chem.* **2014**, *79*, 3747–3751. [CrossRef] [PubMed]
42. Jin, Q.; Jia, G.; Zhang, Y.; Li, C. Modification of supported Pd catalysts by alkalic salts in the selective racemization and dynamic kinetic resolution of primary amines. *Catal. Sci. Technol.* **2014**, *4*, 464–471. [CrossRef]
43. Ma, G.; Xu, Z.; Zhang, P.; Liu, J.; Hao, X.; Ouyang, J.; Liang, P.; You, S.; Jia, X. A novel synthesis of rasagiline via a chemoenzymatic dynamic kinetic resolution. *Org. Process Res. Dev.* **2014**, *18*, 1169–1174. [CrossRef]
44. Mavrynsky, D.; Leino, R. An approach to chemoenzymatic DKR of amines in soxhlet apparatus. *J. Organomet. Chem.* **2014**, *760*, 161–166.
45. De Miranda, A.S.; de Souza, R.O.M.A.; Miranda, L.S.M. Ammonium formate as a green hydrogen source for clean semi-continuous enzymatic dynamic kinetic resolution of (+/−)-α-methylbenzylamine. *RSC Adv.* **2014**, *4*, 13620–13625. [CrossRef]
46. Xia, B.; Cheng, G.; Lin, X.; Wu, Q. Dynamic double kinetic resolution of amines and alcohols under the cocatalysis of Raney nickel/*Candida antarctica* lipase B: From concept to application. *Eur. J. Org. Chem.* **2014**, *2014*, 2917–2923. [CrossRef]
47. Parvulescu, A.N.; Jacobs, P.A.; De Vos, D.E. Heterogeneous raney nickel and cobalt catalysts for racemization and dynamic kinetic resolution of amines. *Adv. Synth. Catal.* **2008**, *350*, 113–121. [CrossRef]
48. Zhang, Y.; Schaufelberger, F.; Sakulsombat, M.; Liu, C.; Ramström, O. Asymmetric synthesis of 1,3-oxathiolan-5-one derivatives through dynamic covalent kinetic resolution. *Tetrahedron* **2014**, *70*, 3826–3831. [CrossRef]
49. Hu, L.; Zhang, Y.; Ramstrom, O. Lipase-catalyzed asymmetric synthesis of oxathiazinanones through dynamic covalent kinetic resolution. *Org. Biomol. Chem.* **2014**, *12*, 3572–3575. [CrossRef] [PubMed]
50. Palo-Nieto, C.; Afewerki, S.; Anderson, M.; Tai, C.-W.; Berglund, P.; Córdova, A. Integrated heterogeneous metal/enzymatic multiple relay catalysis for eco-friendly and asymmetric synthesis. *ACS Catal.* **2016**, *6*, 3932–3940. [CrossRef]
51. Kohler, V.; Turner, N.J. Artificial concurrent catalytic processes involving enzymes. *Chem. Commun.* **2015**, *51*, 450–464. [CrossRef] [PubMed]
52. Groger, H.; Hummel, W. Combining the 'two worlds' of chemocatalysis and biocatalysis towards multi-step one-pot processes in aqueous media. *Curr. Opin. Chem. Biol.* **2014**, *19*, 171–179. [CrossRef] [PubMed]
53. Nirio, M.; Akira, S. Palladium-catalyzed cross-coupling reactions of organoboron compounds. *Chem. Rev.* **1995**, 2457–2483.
54. James, M.T.; Xun-tian, J. The wacker reactions and related alkene oxidation. *Curr. Org. Chem.* **2003**, *7*, 369–396.
55. Costabile, C.; Cavallo, L. Origin of enantioselectivity in the asymmetric Ru-catalyzed metathesis of olefins. *J. Am. Chem. Soc.* **2004**, *126*, 9592–9600. [CrossRef] [PubMed]
56. Jay, C.C.; Deryn, E.F. Ruthenium-catalyzed ring-closing matethesis: Recent advances, limitations and opportunities. *Curr. Org. Chem.* **2006**, *10*, 185–202.
57. Gauchot, V.; Kroutil, W.; Schmitzer, A.R. Highly recyclable chemo-/biocatalyzed cascade reactions with ionic liquids: One-pot synthesis of chiral biaryl alcohols. *Chemistry* **2010**, *16*, 6748–6751. [CrossRef] [PubMed]
58. Boffi, A.; Cacchi, S.; Ceci, P.; Cirilli, R.; Fabrizi, G.; Prastaro, A.; Niembro, S.; Shafir, A.; Vallribera, A. The heck reaction of allylic alcohols catalyzed by palladium nanoparticles in water: Chemoenzymatic synthesis of (R)-(−)-rhododendrol. *ChemCatChem* **2011**, *3*, 347–353. [CrossRef]
59. Schnapperelle, I.; Hummel, W.; Groger, H. Formal asymmetric hydration of non-activated alkenes in aqueous medium through a "chemoenzymatic catalytic system". *Chemistry* **2012**, *18*, 1073–1076. [CrossRef] [PubMed]
60. Denard, C.A.; Huang, H.; Bartlett, M.J.; Lu, L.; Tan, Y.C.; Zhao, H.M.; Hartwig, J.F. Cooperative tandem catalysis by an organometallic complex and a metalloenzyme. *Angew. Chem. Int. Ed.* **2014**, *53*, 465–469. [CrossRef] [PubMed]

61. Denard, C.A.; Bartlett, M.J.; Wang, Y.; Lu, L.; Hartwig, J.F.; Zhao, H. Development of a one-pot tandem reaction combining ruthenium-catalyzed alkene metathesis and enantioselective enzymatic oxidation to produce aryl epoxides. *ACS Catal.* **2015**, *5*, 3817–3822. [CrossRef]

62. Wang, Z.J.; Clary, K.N.; Bergman, R.G.; Raymond, K.N.; Toste, F.D. A supramolecular approach to combining enzymatic and transition metal catalysis. *Nat. Chem.* **2013**, *5*, 100–103. [CrossRef] [PubMed]

63. Kaphan, D.M.; Levin, M.D.; Bergman, R.G.; Raymond, K.N.; Toste, F.D. A supramolecular microenvironment strategy for transition metal catalysis. *Science* **2015**, *350*, 1235–1238. [CrossRef] [PubMed]

64. Heidlindemann, M.; Rulli, G.; Berkessel, A.; Hummel, W.; Gröger, H. Combination of asymmetric organo- and biocatalytic reactions in organic media using immobilized catalysts in different compartments. *ACS Catal.* **2014**, *4*, 1099–1103. [CrossRef]

65. Sperl, J.M.; Carsten, J.M.; Guterl, J.-K.; Lommes, P.; Sieber, V. Reaction design for the compartmented combination of heterogeneous and enzyme catalysis. *ACS Catal.* **2016**, *6*, 6329–6334. [CrossRef]

66. Metzner, R.; Hummel, W.; Wetterich, F.; König, B.; Gröger, H. Integrated biocatalysis in multistep drug synthesis without intermediate isolation: A de novo approach toward a rosuvastatin key building block. *Org. Process Res. Dev.* **2015**, *19*, 635–638. [CrossRef]

67. Sato, H.; Hummel, W.; Groger, H. Cooperative catalysis of noncompatible catalysts through compartmentalization: Wacker oxidation and enzymatic reduction in a one-pot process in aqueous media. *Angew. Chem. Int. Ed.* **2015**, *54*, 4488–4492. [CrossRef] [PubMed]

68. Gómez Baraibar, Á.; Reichert, D.; Mügge, C.; Seger, S.; Gröger, H.; Kourist, R. A one-pot cascade reaction combining an encapsulated decarboxylase with a metathesis catalyst for the synthesis of bio-based antioxidants. *Angew. Chem. Int. Ed.* **2016**, *55*, 14823–14827. [CrossRef] [PubMed]

69. Lang, X.; Zhao, J.; Chen, X. Cooperative photoredox catalysis. *Chem. Soc. Rev.* **2016**, *45*, 3026–3038. [CrossRef] [PubMed]

70. Tran, N.H.; Nguyen, D.; Dwaraknath, S.; Mahadevan, S.; Chavez, G.; Nguyen, A.; Dao, T.; Mullen, S.; Nguyen, T.A.; Cheruzel, L.E. An efficient light-driven P450 BM3 biocatalyst. *J. Am. Chem. Soc.* **2013**, *135*, 14484–14487. [CrossRef] [PubMed]

71. Park, J.H.; Lee, S.H.; Cha, G.S.; Choi, D.S.; Nam, D.H.; Lee, J.H.; Lee, J.K.; Yun, C.H.; Jeong, K.J.; Park, C.B. Cofactor-free light-driven whole-cell cytochrome P450 catalysis. *Angew. Chem. Int. Ed.* **2015**, *54*, 969–973. [CrossRef] [PubMed]

72. Choudhury, S.; Baeg, J.O.; Park, N.J.; Yadav, R.K. A photocatalyst/enzyme couple that uses solar energy in the asymmetric reduction of acetophenones. *Angew. Chem. Int. Ed.* **2012**, *51*, 11624–11628. [CrossRef] [PubMed]

73. Lee, H.Y.; Ryu, J.; Kim, J.H.; Lee, S.H.; Park, C.B. Biocatalyzed artificial photosynthesis by hydrogen-terminated silicon nanowires. *ChemSusChem* **2012**, *5*, 2129–2132. [CrossRef] [PubMed]

74. Mifsud, M.; Gargiulo, S.; Iborra, S.; Arends, I.W.; Hollmann, F.; Corma, A. Photobiocatalytic chemistry of oxidoreductases using water as the electron donor. *Nat. Commun.* **2014**, *5*, 3145. [CrossRef] [PubMed]

75. Peers, M.K.; Toogood, H.S.; Heyes, D.J.; Mansell, D.; Coe, B.J.; Scrutton, N.S. Light-driven biocatalytic reduction of α,β-unsaturated compounds by ene reductases employing transition metal complexes as photosensitizers. *Catal. Sci. Technol.* **2016**, *6*, 169–177. [CrossRef] [PubMed]

76. Heinisch, T.; Ward, T.R. Artificial metalloenzymes based on the biotin-streptavidin technology: Challenges and opportunities. *Acc. Chem. Res.* **2016**, *49*, 1711–1721. [CrossRef] [PubMed]

77. Rosati, F.; Roelfes, G. Artificial metalloenzymes. *ChemCatChem* **2010**, *2*, 916–927. [CrossRef]

78. Nastri, F.; Chino, M.; Maglio, O.; Bhagi-Damodaran, A.; Lu, Y.; Lombardi, A. Design and engineering of artificial oxygen-activating metalloenzymes. *Chem. Soc. Rev.* **2016**, *45*, 5020–5054. [CrossRef] [PubMed]

79. Petrik, I.D.; Liu, J.; Lu, Y. Metalloenzyme design and engineering through strategic modifications of native protein scaffolds. *Curr. Opin. Chem. Biol.* **2014**, *19*, 67–75. [CrossRef] [PubMed]

80. Bos, J.; Roelfes, G. Artificial metalloenzymes for enantioselective catalysis. *Curr. Opin. Chem. Biol.* **2014**, *19*, 135–143. [CrossRef] [PubMed]

81. Yu, F.; Cangelosi, V.M.; Zastrow, M.L.; Tegoni, M.; Plegaria, J.S.; Tebo, A.G.; Mocny, C.S.; Ruckthong, L.; Qayyum, H.; Pecoraro, V.L. Protein design: Toward functional metalloenzymes. *Chem. Rev.* **2014**, *114*, 3495–3578. [CrossRef] [PubMed]

82. Drienovská, I.; Rioz-Martínez, A.; Draksharapu, A.; Roelfes, G. Novel artificial metalloenzymes by in vivo incorporation of metal-binding unnatural amino acids. *Chem. Sci.* **2015**, *6*, 770–776. [CrossRef]

83. Heinisch, T.; Pellizzoni, M.; Durrenberger, M.; Tinberg, C.E.; Kohler, V.; Klehr, J.; Haussinger, D.; Baker, D.; Ward, T.R. Improving the catalytic performance of an artificial metalloenzyme by computational design. *J. Am. Chem. Soc.* **2015**, *137*, 10414–10419. [CrossRef] [PubMed]

84. Lewis, J.C. Metallopeptide catalysts and artificial metalloenzymes containing unnatural amino acids. *Curr. Opin. Chem. Biol.* **2015**, *25*, 27–35. [CrossRef] [PubMed]

85. Key, H.M.; Dydio, P.; Clark, D.S.; Hartwig, J.F. Abiological catalysis by artificial haem proteins containing noble metals in place of iron. *Nature* **2016**, *534*, 534–537. [CrossRef] [PubMed]

86. Jeschek, M.; Reuter, R.; Heinisch, T.; Trindler, C.; Klehr, J.; Panke, S.; Ward, T.R. Directed evolution of artificial metalloenzymes for in vivo metathesis. *Nature* **2016**, *537*, 661–665. [CrossRef] [PubMed]

87. Wilson, Y.M.; Durrenberger, M.; Nogueira, E.S.; Ward, T.R. Neutralizing the detrimental effect of glutathione on precious metal catalysts. *J. Am. Chem. Soc.* **2014**, *136*, 8928–8932. [CrossRef] [PubMed]

88. Kohler, V.; Wilson, Y.M.; Durrenberger, M.; Ghislieri, D.; Churakova, E.; Quinto, T.; Knorr, L.; Haussinger, D.; Hollmann, F.; Turner, N.J.; et al. Synthetic cascades are enabled by combining biocatalysts with artificial metalloenzymes. *Nat. Chem.* **2013**, *5*, 93–99. [CrossRef] [PubMed]

89. Lloyd, J.R. Microbial reduction of metals and radionuclides. *FEMS Microbiol. Rev.* **2003**, *27*, 411–425. [CrossRef]

90. De Windt, W.; Aelterman, P.; Verstraete, W. Bioreductive deposition of palladium (0) nanoparticles on shewanella oneidensis with catalytic activity towards reductive dechlorination of polychlorinated biphenyls. *Environ. Microbiol.* **2005**, *7*, 314–325. [CrossRef] [PubMed]

91. Raab, R.M.; Tyo, K.; Stephanopoulos, G. Metabolic engineering. *Adv. Biochem. Eng. Biotechnol.* **2005**, *100*, 1–17. [PubMed]

92. Foulkes, J.M.; Malone, K.J.; Coker, V.S.; Turner, N.J.; Lloyd, J.R. Engineering a biometallic whole cell catalyst for enantioselective deracemization reactions. *ACS Catal.* **2011**, *1*, 1589–1594. [CrossRef]

93. Royer, G.P.; Chow, W.-S.; Hatton, K.S. Palladium/polyethylenimine catalysts. *J. Mol. Catal.* **1985**, *31*, 1–13. [CrossRef]

94. Sirasani, G.; Tong, L.; Balskus, E.P. A biocompatible alkene hydrogenation merges organic synthesis with microbial metabolism. *Angew. Chem. Int. Ed.* **2014**, *53*, 7785–7788. [CrossRef] [PubMed]

95. Yang, M.; Li, J.; Chen, P.R. Transition metal-mediated bioorthogonal protein chemistry in living cells. *Chem. Soc. Rev.* **2014**, *43*, 6511–6526. [CrossRef] [PubMed]

96. Volker, T.; Dempwolff, F.; Graumann, P.L.; Meggers, E. Progress towards bioorthogonal catalysis with organometallic compounds. *Angew. Chem. Int. Ed.* **2014**, *53*, 10536–10540. [CrossRef] [PubMed]

97. Tomás-Gamasa, M.; Martínez-Calvo, M.; Couceiro, J.R.; Mascareñas, J.L. Transition metal catalysis in the mitochondria of living cells. *Nat. Commun.* **2016**, *7*, 12538. [CrossRef] [PubMed]

98. Li, J.; Yu, J.; Zhao, J.; Wang, J.; Zheng, S.; Lin, S.; Chen, L.; Yang, M.; Jia, S.; Zhang, X.; et al. Palladium-triggered deprotection chemistry for protein activation in living cells. *Nat. Chem.* **2014**, *6*, 352–361. [CrossRef] [PubMed]

99. Burtscher, D.; Grela, K. Aqueous olefin metathesis. *Angew. Chem. Int. Ed.* **2009**, *48*, 442–454. [CrossRef] [PubMed]

100. Binder, J.B.; Raines, R.T. Olefin metathesis for chemical biology. *Curr. Opin. Chem. Biol.* **2008**, *12*, 767–773. [CrossRef] [PubMed]

101. Gulajski, L.; Michrowska, A.; Naroznik, J.; Kaczmarska, Z.; Rupnicki, L.; Grela, K. A highly active aqueous olefin metathesis catalyst bearing a quaternary ammonium group. *ChemSusChem* **2008**, *1*, 103–109. [CrossRef] [PubMed]

102. Schaper, L.A.; Hock, S.J.; Herrmann, W.A.; Kuhn, F.E. Synthesis and application of water-soluble NHC transition-metal complexes. *Angew. Chem. Int. Ed.* **2013**, *52*, 270–289. [CrossRef] [PubMed]

103. Chaturvedi, D.; Barua, N.C. Recent developments on carbon-carbon bond forming reactions in water. *Curr. Org. Synth.* **2012**, *9*, 17–30. [CrossRef]

104. Datta, S.; Christena, L.R.; Rajaram, Y.R.S. Enzyme immobilization: An overview on techniques and support materials. *3 Biotech* **2013**, *3*, 1–9. [CrossRef]

105. Engström, K.; Johnston, E.V.; Verho, O.; Gustafson, K.P.J.; Shakeri, M.; Tai, C.-W.; Bäckvall, J.-E. Co-immobilization of an enzyme and a metal into the compartments of mesoporous silica for cooperative tandem catalysis: An artificial metalloenzyme. *Angew. Chem. Int. Ed.* **2013**, *52*, 14006–14010. [CrossRef] [PubMed]

106. Engelmark Cassimjee, K.; Kadow, M.; Wikmark, Y.; Svedendahl Humble, M.; Rothstein, M.L.; Rothstein, D.M.; Backvall, J.E. A general protein purification and immobilization method on controlled porosity glass: Biocatalytic applications. *Chem. Commun.* **2014**, *50*, 9134–9137. [CrossRef] [PubMed]

107. Sheldon, R.A. Cross-linked enzyme aggregates (CLEA®s): Stable and recyclable biocatalysts. *Biochem. Soc. Trans.* **2007**, *35*, 1583–1587. [CrossRef] [PubMed]

108. Frushicheva, M.P.; Mills, M.J.; Schopf, P.; Singh, M.K.; Prasad, R.B.; Warshel, A. Computer aided enzyme design and catalytic concepts. *Curr. Opin. Chem. Biol.* **2014**, *21*, 56–62. [CrossRef] [PubMed]

109. Porebski, B.T.; Buckle, A.M. Consensus protein design. *Protein Eng. Des. Sel.* **2016**, *29*, 245–251. [CrossRef] [PubMed]

110. Packer, M.S.; Liu, D.R. Methods for the directed evolution of proteins. *Nat. Rev. Genet.* **2015**, *16*, 379–394. [CrossRef] [PubMed]

111. Bistri, O.; Reinaud, O. Supramolecular control of transition metal complexes in water by a hydrophobic cavity: A bio-inspired strategy. *Org. Biomol. Chem.* **2015**, *13*, 2849–2865. [CrossRef] [PubMed]

112. Li, A.; Ilie, A.; Sun, Z.; Lonsdale, R.; Xu, J.H.; Reetz, M.T. Whole-cell-catalyzed multiple regio- and stereoselective functionalizations in cascade reactions enabled by directed evolution. *Angew. Chem. Int. Ed.* **2016**, *55*, 12026–12029. [CrossRef] [PubMed]

113. Both, P.; Busch, H.; Kelly, P.P.; Mutti, F.G.; Turner, N.J.; Flitsch, S.L. Whole-cell biocatalysts for stereoselective C–H amination reactions. *Angew. Chem. Int. Ed.* **2016**, *55*, 1511–1513. [CrossRef] [PubMed]

*Review*

# Old Yellow Enzyme-Catalysed Asymmetric Hydrogenation: Linking Family Roots with Improved Catalysis

Anika Scholtissek [1], Dirk Tischler [1], Adrie H. Westphal [2], Willem J. H. van Berkel [2] and Caroline E. Paul [3,*]

1   Interdisciplinary Ecological Center, Institute of Biosciences, Environmental Microbiology Group, Technical University Bergakademie Freiberg, 09599 Freiberg, Germany; anika.scholtissek@gmail.com (A.S.); dirk-tischler@email.de (D.T.)
2   Laboratory of Biochemistry, Wageningen University & Research, Stippeneng 4, 6708 WE Wageningen, The Netherlands; adrie.westphal@wur.nl (A.H.W.); willem.vanberkel@wur.nl (W.J.H.v.B.)
3   Department of Biotechnology, Delft University of Technology, Van der Maasweg 9, 2629 Delft, The Netherlands
*   Correspondence: c.e.paul@tudelft.nl; Tel.: +31-1527-84616

Academic Editors: Cesar Mateo and Jose M. Palomo
Received: 16 March 2017; Accepted: 25 April 2017; Published: 29 April 2017

**Abstract:** Asymmetric hydrogenation of activated alkenes catalysed by ene-reductases from the old yellow enzyme family (OYEs) leading to chiral products is of potential interest for industrial processes. OYEs' dependency on the pyridine nucleotide coenzyme can be circumvented through established artificial hydride donors such as nicotinamide coenzyme biomimetics (NCBs). Several OYEs were found to exhibit higher reduction rates with NCBs. In this review, we describe a new classification of OYEs into three main classes by phylogenetic and structural analysis of characterized OYEs. The family roots are linked with their use as chiral catalysts and their mode of action with NCBs. The link between bioinformatics (sequence analysis), biochemistry (structure–function analysis), and biocatalysis (conversion, enantioselectivity and kinetics) can enable an early classification of a putative ene-reductase and therefore the indication of the binding mode of various activated alkenes.

**Keywords:** old yellow enzymes; nicotinamide coenzyme biomimetics; cofactor analogues; classification of OYE; oxidoreductases; asymmetric hydrogenation; selective reduction; phylogenetics

## 1. Introduction

The 2001 Nobel prize in chemistry awarded to William S. Knowles and Ryōji Noyori internationally highlighted the importance of catalysed asymmetric hydrogenation reactions [1]. Particularly, the creation of one to two chiral centres through asymmetric hydrogenation of C=C bonds is a highly valuable reaction in organic synthesis [2]. Common synthetic routes for *cis*-hydrogenation are accomplished via homogeneous chiral catalysts composed of precious metals such as rhodium (Rh), ruthenium (Ru) or iridium (Ir), and phosphine ligands such as chiral mono- and di-phosphines, $C_2$-symmetric bisoxazoline ligands or $C_2$-symmetric *N*-heterocyclic carbenes, respectively [3]. In comparison, synthetic methods for asymmetric *trans*-hydrogenation to afford the stereo-complementary products are scarce [4].

A highly competitive tool for asymmetric *trans*-hydrogenation is the biocatalytic route using ene-reductases (ERs) of the old yellow enzyme family (OYEs, EC 1.6.99.1). The performance of these enzymes is of potential interest for industrial processes due to their high regio-, stereo- and enantioselectivity, and an expanding substrate scope [5–9]. The substrate spectrum of OYEs includes activated alkenes with an electron withdrawing group (EWG) such as aldehyde, ketone, anhydride [8,10,11], nitro [9,12,13], (di)ester [8,14–17], (di)carboxylic acid [18–20], cyclic imide [21–23], nitrile [24], β-cyanoacrylate [25], β-nitroacrylate [26], and several other functional groups [27]. There are many examples of the high industrial potential of OYEs for the synthesis of valuable target products [28–31]. YqjM was found to produce enantiomerically pure (*R*)-profens and is applied in the synthesis of (*R*)-flurbiprofen methyl ester [32]. Flurbiprofen belongs to the non-steroidal anti-inflammatory drugs (NSAIDs) and is used at the appearance of dental pain or sore throat. A library of OYEs was used for the asymmetric reduction of β-cyanoacrylate esters to yield a precursor of pregabalin, an anticonvulsant for epilepsy or fibromyalgia [33]. A similar OYE library reduced α-, β- and γ-substituted α,β-unsaturated butyrolactones [34], structural components of macrocyclic antibiotics [34,35]. A valuable overview for OYE-catalysed reactions from recent studies has been compiled by Toogood and co-workers, including an extensive substrate profile of isolated OYEs [36].

OYEs are flavin mononucleotide (FMN)-containing ERs and catalyse the selective asymmetric reduction of activated C=C bonds at the expense of the pyridine nucleotide coenzyme NAD(P)H, following a bi-bi ping-pong kinetic mechanism (Scheme 1). In the reductive half-reaction, FMN is reduced through hydride transfer from NAD(P)H (C4) [37]. In the oxidative half-reaction, a hydride is transferred from the N5-atom of the reduced flavin to the Cβ-atom of the activated alkene. A tyrosine residue provides a proton to the Cα-atom, thus completing the reduction of the C=C bond [9,37,38]. This mechanism leads to an *anti*-addition (*trans*-fashion) hydrogenation and is supported by recent quantum mechanics/molecular mechanics calculations [39]. The reductive half-reaction was experimentally investigated in detail for OYE1 by Massey and co-workers [40]. Binding of NADPH to the oxidised enzyme-FMN complex led to the observation of a transient concentration-dependent Michaelis complex. After NADPH binding, generation of the reduced enzyme-NADP⁺ complex was noticed as a long wavelength absorbance band. Formation of this charge-transfer complex indicated that the electron and subsequent hydride transfer requires π–π stacking between the pyridinium ring of the nicotinamide cofactor and the isoalloxazine ring of the FMN [40,41].

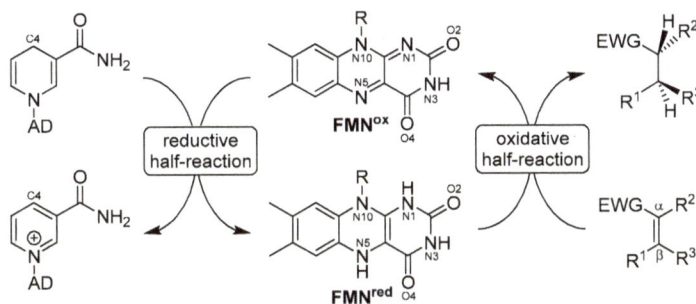

**Scheme 1.** OYE-catalysed asymmetric hydrogenation of activated alkenes through a bi-bi ping-pong mechanism producing one to two chiral centres. AD = adenine dinucleotide; R = ribose phosphate; EWG = electron withdrawing group.

Although NADPH is the preferred physiological coenzyme for OYE, the dependency on this commercially expensive compound can be circumvented by established recycling systems with dehydrogenases [12,18,42–45], with alternative sources of hydride [46,47], or through a nicotinamide-independent disproportionation coupling reaction [48–51]. A highly promising and elegant alternative is the use of relatively inexpensive nicotinamide coenzyme biomimetics (NCBs) [52–56]. The latter compounds retain the pyridine ring structure, substituted with varied functional groups either on the N1 nitrogen (NCBs **1–2**, **6–7**, Figure 1) or at the C3 carbon (NCBs **3–5**) [55]. As with the natural coenzyme, the correct positioning of the pyridine ring in the active site is crucial for optimal hydride transfer [52,57].

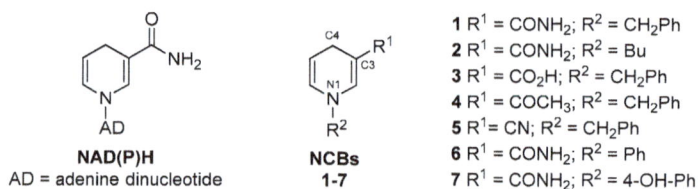

1 $R^1$ = CONH$_2$; $R^2$ = CH$_2$Ph
2 $R^1$ = CONH$_2$; $R^2$ = Bu
3 $R^1$ = CO$_2$H; $R^2$ = CH$_2$Ph
4 $R^1$ = COCH$_3$; $R^2$ = CH$_2$Ph
5 $R^1$ = CN; $R^2$ = CH$_2$Ph
6 $R^1$ = CONH$_2$; $R^2$ = Ph
7 $R^1$ = CONH$_2$; $R^2$ = 4-OH-Ph

**NAD(P)H**
AD = adenine dinucleotide

**NCBs**
1-7

**Figure 1.** Nicotinamide coenzyme biomimetics (NCBs) previously used in OYE-catalysed hydrogenations to replace NAD(P)H [52,55,58].

Since 2013, several OYEs were found to exhibit high catalytic activities with different NCB analogues [52,57–59]. In a first study, YqjM, TsOYE and RmOYE were screened against NCBs **1–5** (Figure 1) [52], followed by more extensive kinetic studies with a panel of OYEs [59]. Depending on the applied NCBs, OYEs differ in their reaction rates and the catalytic efficiency ($k_{cat}/K_m$) is at times even higher with NCBs than with the natural coenzyme, as discussed further in Section 4 [22,58,59].

In this review, we suggest linking the evolutionary history of OYEs with their activity with NCBs and their use as chiral catalysts on various substrates. To this end, we start with a phylogenetic classification of OYEs characterised thus far, and relate this classification to their structural and biocatalytic properties.

## 2. Phylogenetic Classification of OYEs

OYEs are ubiquitous in Nature [21,60]. Many ERs from the OYE family have been (recombinantly) expressed and characterised over the last 25 years. Currently, we have access to approximately 63 characterised and "ready-to-use" OYEs from plantae, fungi and bacteria. Tables 1 and 2 indicate the distribution of those well-characterised OYEs, with respect to their domain eukaryota (Table 1) and prokaryota (Table 2).

One third of the characterised OYEs have been obtained from eukaryota, mainly from the kingdom fungi, subkingdom of dikarya, phyla ascomyceta. However, the fungal OYEs originate from different families such as *Saccharomycotina* [61,62], and *Pezizomycotina* [63,64]. Fewer studies have been performed on OYEs originating from plants. Nevertheless, the enzymes AtOPR1–AtOPR3 from *Arabidopsis thaliana* [65,66], and LeOPR1–LeOPR3 from *Solanum lycopersicum* (tomato) [67,68], were characterised according to their structure, function and physiological role. Two thirds of the characterised OYEs have been obtained from various classes of bacteria including proteobacteria (28%) [69,70], actinobacteria (5%) [22,71,72], bacteroidetes (5%) [73,74], firmicutes (10%) [75–77], deinococcus-thermus (3%) [78–80], and cyanobacteria (17%) [21,81].

The bacterial OYEs investigated until now have been categorised in classical and thermophilic-like (formerly YqjM-like) enzymes [9]. In 2016, Nizam and co-workers performed a comprehensive study of 424 putative OYEs from 60 fungal species and indicated a novel group among fungal OYEs [63]. However, none of these enzymes has been characterised thus far.

The first classical OYE (OYE1) was isolated from brewers' bottom yeast (*Saccharomyces carlsbergensis*) in 1932 [5,82]. The same protein was the basis for the first OYE crystal structure, uncovering a TIM-barrel topology, related to trimethylamine dehydrogenase [83]. Since then, many classical OYEs were identified from proteobacteria (NCR [84], MR [85], PETNR [8]), flavobacteria (Chr-OYE2 [74]), cyanobacteria [21], yeasts (OYE1–OYE3 [86,87], CYE [62]) and plants [67], outlined in Tables 1 and 2.

**Table 1.** Sources of biochemically characterised eukaryotic OYEs.

| Kingdom | Enzyme (Accession Number) | Source | Reference(s) |
|---|---|---|---|
| Fungi | OYE1 (CAA37666) | *Saccharomyces pastorianus* | [61] |
| | OYE2 (AAA83386) | *Saccharomyces cerevisiae* | [87] |
| | OYE3 (AAA64522) | *Saccharomyces cerevisiae* | [88] |
| | EBP1 (AAA18013) | *Candida albicans* | [89] |
| | HYE1 (AAN09952) | *Ogataea angusta* | [90] |
| | HYE2 (AAN09953) | *Ogataea angusta* | [90] |
| | CYE (BAD24850) | *Kluyveromyces marxianus* | [62,91] |
| | KYE1 (AAA98815) | *Kluyveromyces lactis* | [6,92] |
| | OYE2.6 (ABN66026) | *Scheffersomyces stipitis* CBS 6054 | [93,94] |
| | ArOYE1–3 (AHL17019, AHL1720, AHL17021) | *Ascochyta rabiei* | [64] |
| | ClER (EEQ40235) | *Clavispora lusitaniae* ATCC 42720 | [95] |
| | MgER (EDK41665) | *Meyerozyma guilliermondii* ATCC 6260 | [96] |
| Plants | LeOPR1 (NP_001234781) | *Solanum lycopersicum* | [68] |
| | LeOPR2 (NP_001233868) | *Solanum lycopersicum* | [67] |
| | LeOPR3 (NP_001233873) | *Solanum lycopersicum* | [67] |
| | AtOPR1 (NP_177794) | *Arabidopsis thaliana* | [65] |
| | AtOPR2 (NP_177795) | *Arabidopsis thaliana* | [97] |
| | AtOPR3 (NP_001077884) | *Arabidopsis thaliana* | [66] |

Colours are assigned based on a new classification according to the dendrogram in Figure 2. Yellow (class I) contains classical OYEs originating from plants. Grey (class II) contains OYE homologues originating from fungal species.

**Table 2.** Sources of biochemically characterized prokaryotic OYEs.

| Group/Order | Enzyme (Ncbi Accession) | Source | Reference(s) |
|---|---|---|---|
| Proteobacteria/ α-Proteobacteria | NerA/GTNR (CAA74280) | *Agrobacterium radiobacter* | [98] |
| | NCR (AAV90509) | *Zymomonas mobilis* | [84] |
| | GluER (AAW60280) | *Gluconobacter oxidans* DSM 2343 | [99] |
| Proteobacteria/ β-Proteobacteria | FOYE-1 (KRH78075) | *Ferrovum* sp. JA12 | [23] |
| | RmER (ABF11721) | *Cupriavidus metallidurans* CH34 | [80] |
| | Achr-OYE3 (AFK73187) | *Achromobacter* sp. JA81 | [16] |
| | Achr-OYE4 (AFK73188) | *Achromobacter* sp. JA81 | [16,17] |
| Proteobacteria/ γ-Proteobacteria | MR (AAC43569) | *Pseudomonas putida* M10 | [85] |
| | PETNR (AAB38683) | *Enterobacter cloacae* PB2 | [69] |
| | NemR/NemA (BAA13186) | *Escherichia coli* | [100] |
| | NemA2 (AHC69715) | *Pseudomonas putida* ATCC 17453 | [101] |
| | XenA (AAF02538) | *Pseudomonas putida* II-B | [102] |
| | XenA2 (AHH54488) | *Pseudomonas putida* ATCC 17453 | [101] |
| | XenB (AAF02539) | *Pseudomonas fluorescens* I-C | [102] |
| | XenB2 (AGS77941) | *Pseudomonas putida* ATCC 17453 | [101] |
| | YersER (WP_032896199) | *Yersinia bercovieri* | [6] |
| | SYE1 (AAN55488) | *Shewanella oneidensis* | [103] |
| | SYE3 (AAN57126) | *Shewanella oneidensis* | [103] |
| | SYE4 (AAN56390) | *Shewanella oneidensis* | [103] |
| Actinobacteria | OYERo2 (ALL54975) | *Rhodococcus opacus* 1CP | [22] |
| | Nox (ALG03744) | *Rhodococcus erythropolis* | [72] |
| | PfvC (AFF18622) | *Arthrobacter* sp. JBH1 | [71] |
| Bacteroidetes/ Flavobacteria | Chr-OYE1 (ALE60336) | *Chryseobacterium* sp. CA49 | [73] |
| | Chr-OYE2 (ALE60337) | *Chryseobacterium* sp. CA49 | [73] |
| | Chr-OYE3 (AHV90721) | *Chryseobacterium* sp. CA49 | [74] |
| Firmicutes/(Bacilli, Clostridia) | YqjM (BAA12619) | *Bacillus subtilis* strain 168 | [75] |
| | YqiG (BAA12582) | *Bacillus subtilis* strain 168 | [104] |
| | GkOYE (BAD76617) | *Geobacillus kaustophilus* DSM7263 | [76] |
| | GeoER (BAO37313) | *Geobacillus* sp. 30 | [105] |
| | LacER (ADK19581) | *Lactobacillus casei* str. Zhang | [10] |
| | TOYE (ABY93685) | *Thermoanaerobacter pseudethanolicus* E39 | [77] |

**Table 2.** *Cont.*

| Group/Order | Enzyme (Ncbi Accession) | Source | Reference(s) |
|---|---|---|---|
| Deinococcus-Thermus | TsOYE (CAP16804) | *Thermus scotoductus* SA-01 | [79] |
| | DrER (AAF11740) | *Deinococcus radiodurans* R1 | [80] |
| Cyanobacteria/ (Gloebacteria, Oscillatoriophycidea, Nostocales) | GloeoER (BAC91769) | *Gloeobacter violaceus* PCC7421 | [81] |
| | CyanothER1 (ACK64210) | *Cyanothece* sp. PCC 8801 | [81] |
| | CyanothER2 (ACK65723) | *Cyanothece* sp. PCC 8801 | [81] |
| | LyngbyaER1 (EAW37813) | *Lyngbya* sp. PCC 8106 | [81] |
| | AcaryoER1 (ABW29811) | *Acaryochloris marina* MBIC11017 | [81] |
| | AcaryoER3 (ABW32756) | *Acaryochloris marina* MBIC11017 | [81] |
| | SynER (ABB56505) | *Synechococcus elongatus* PCC 7942 | [21] |
| | NospuncER1 (ACC84535) | *Nostoc punctiforme* PCC 73102 | [81] |
| | NostocER1 (BAB73564) | *Nostoc* sp. PCC 7120 | [81] |
| | AnabaenaER3 (ABA25236) | *Anabaena variabilis* ATCC 29413 | [81] |

Colours are assigned based on a new classification according to the dendrogram in Figure 2. Yellow (class I) contains classical OYEs originating from bacteria. Green (class III) contains thermophilic-like OYE homologues originating from bacteria. Non-highlighted OYEs could not be assigned to classes I–III.

The discovery of a second OYE subclass twelve years ago led Macheroux and co-workers to publish the structure of YqjM, an OYE from the *Bacillus subtilis* strain 168 [75,106]. In contrast to all other OYEs known at the time, YqjM exhibited unique structural properties, such as its occurrence as a homotetramer, and the presence of an arginine at the C-terminus involved in the substrate binding of the adjacent monomer. Next, the thermostable TsOYE (formerly CrS) and TOYE were isolated from *Thermus scotoductus* SA-01 and *Thermoanaerobacter pseudethanolicus*, respectively [77–79]. Based on a sequence alignment of known and putative OYE homologues from mesophilic and thermophilic organisms, the renaming of the "YqjM" subclass into the "thermophilic-like" was proposed by Scrutton and co-workers in 2010 [77–79]. Subsequently, the number of characterised OYEs in this subclass has risen to sixteen (Table 2, highlighted in green).

OYEs from the thermophilic-like class possess shorter protein sequences (between 337 and 371 amino acid residues) than classical OYEs (between 349 and 412 amino acid residues). Thermophilic-like OYEs have an average increased thermal stability compared to classical ones. High melting temperatures were observed for TsOYE ($T_{opt}$ = 65 °C [79]), TOYE ($T_m$ > 70 °C [77]), GkOYE ($T_m$ = 76–82 °C [76]), GeoER ($T_{opt}$ = 70 °C [105]), and FOYE-1 ($T_{opt}$ = 50 °C [23]). The thermostability of TsOYE and TOYE was assigned to a high proline content within loops and turns (typical for *Thermus* species) as well as to strong inter-subunit interactions through hydrogen bonding and complex salt bridge networks at the dimerization interface [77,78].

We recently described FOYE-1 as a thermostable OYE [23]. Surprisingly, FOYE-1 showed highest sequence identity (55% and 50%), and therefore closest phylogenetical relationship, to the mesophilic counterparts RmER and DrER [23]. DrER and RmER are not an exception with respect to the non-thermostable OYE relatives YqjM, XenA, and OYERo2, all clustering in the thermophilic-like subclass. These OYEs have a proline content below 7%, and are mostly stabilized through single salt bridges. Due to these varieties in the thermophilic-like subclass, we suggest the classification be updated through a phylogenetic analysis of all biochemically characterized OYEs (Figure 2).

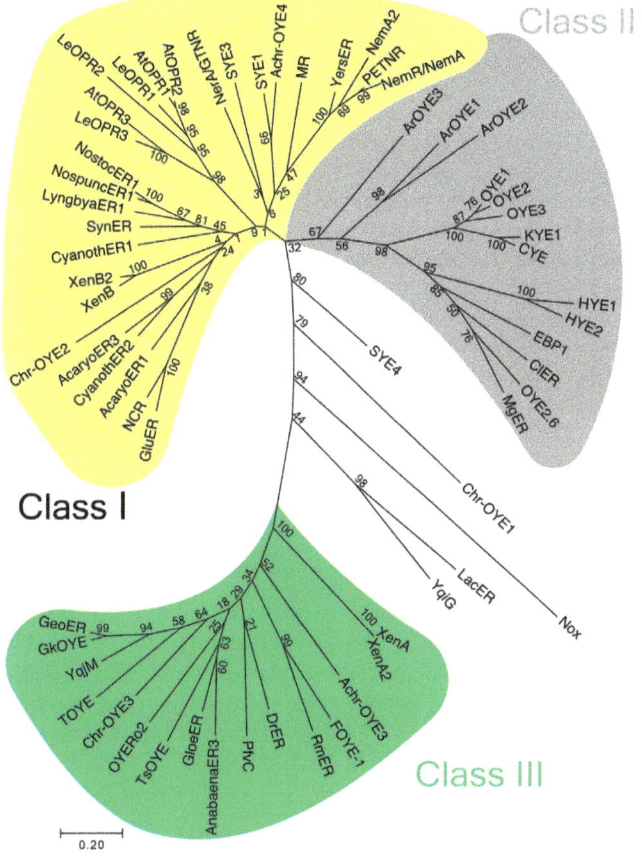

**Figure 2.** Dendrogram showing the relationship of 63 characterised OYEs from plants, fungi and bacteria. Corresponding accession numbers and sources are given in Tables 1 and 2. Class I (yellow) contains classical OYEs originating from plants and bacteria. Class II (grey) contains classical OYEs originating only from fungal species. Class III (green) contains the thermophilic-like and mesophilic OYEs originating from various bacteria. The maximum likelihood distance tree (Mega7-mac computed) was calculated with replications of 500 bootstraps. Corresponding values are shown as nodes at all branches. The corresponding alignment was produced via ClustalW alignment applying the GONNET protein weight matrix.

Calculating the phylogenetic distance tree of 63 characterised OYEs revealed three instead of two comprehensive branches (Figure 2). Branch 1 (highlighted in yellow) contains many classical OYEs from plants, actino- and proteobacteria, flavobacteria, but also another distinct subclade from cyanobacteria. We designate this branch "class I". Branch 2 ("class II", highlighted in grey) appears closely related to branch 1 and contains exclusively classical enzymes from fungi. Branch 3 ("class III", highlighted in green) is further away from branch 1 and 2 and contains–like class I–not only bacterial OYEs from actino-, proteo- and cyanobacteria, but also OYEs from deinococci, flavobacteria and firmicutes. This branch contains the traditional thermophilic-like OYEs. Furthermore, several sequences (SYE4, Chr-OYE1, Nox, LacER and YqiG) occurred in a position where they cannot be assigned yet. They may represent evolutionary intermediates/predecessors of class I and class III enzymes as they are of bacterial origin. The short distance between class I and class II hints towards a close relatedness and therefore a co-driven evolution. Our phylogenetic analysis allowed no distinction between thermostable and non-thermostable OYEs within class III since thermostability does not display a logical pattern.

Among conventional phylogeny methods, we considered the amino acid composition of all analysed OYEs regarding their early/simple (particularly Ala, Thr, and Val) and late/sophisticated (particularly Cys, Leu, Phe, Trp, and Tyr) amino acid residues [107,108]. The quotient of late over early amino acids was used to determine a comparable time of evolvement factor. Therefore, a low number represents a primary chain mostly containing early/simple amino acids. Regarding the OYEs from class I without the subclade from cyanobacteria, an average factor of 0.76 was determined. By contrast, class II's average factor amounted to 0.94. This result would indicate a successive evolution during which class II evolved from class I. However, the more remote class III could have evolved in a convergent evolution with class I since the average factor was similar (0.77). The convergent evolution hypothesis is strengthened by the difference in average amino acid length as well as by the biochemical and structural properties, which both differ among class I and class III. OYEs from cyanobacteria (included in class I) seem also to be recruited later due to the average factor of 0.88. To intensify this finding, main properties of class I, II and III regarding their sequences and structures are discussed in the following section dealing with the structural classification of those three classes. Furthermore, this information enables to subclassify class III OYEs and to assign the lost proteins SYE4, Chr-OYE1, Nox, LacER and YqiG to one of the three classes or to confirm their independent status, respectively.

## 3. Structural Classification of OYEs: From Sequence to Structure

OYE homologues show high conservation of amino acids involved in the binding of flavin, substrates and/or inhibitors, as well as for those involved in the formation of the dimeric interface. However, there are many differences in the conservation of distinct residues depending on the class of OYEs mentioned above. A multiple sequence alignment was performed with five representative sequences from each class (Figure S1). This section compares the sequences of OYE classes I–III enzymes and discusses the significant effects on the structure of OYEs.

### 3.1. Multiple Sequence Alignment

Approximately 15% (56 amino acids (aa)) are OYE-conserved residues in all three classes (Figure S1, highlighted in black). Moreover, 40 additional aa are conserved especially in class I (highlighted in yellow), 32 aa in class II (highlighted in grey) and 43 aa in class III (highlighted in green). Evidently, class I and class II share significantly more conserved residues when compared to class III, resulting in the minor distance (Figure 2).

The alignment including the five non-assignable sequences also allows matching of these OYE enzymes with the defined classes. The sequences SYE-4 and Chr-OYE1 are more closely related to class I since they share 63% and 45% of the class I conserved residues but only 34% (25%) for class II and 9% (21%) for class III, respectively. Nox and YqiG are more related to class III, sharing 37% and 40% of class III conserved residues whereas they jointly own only 23% (34%) of class I, and 25% (19%) of class II conserved residues. LacER shares 39% (class I), 19% (class II) and 30% (class III) and is somewhere in between. Interestingly, all five proteins do not possess the Cys26 and the Arg336 (YqjM numbering), which are highly conserved in class III and are involved in flavin and substrate binding. The alignment shows they contain many general OYE motifs. However, all of them comprise motifs from class I/II but also from class III generating a reasonable alignment in between of classes I–III. Therefore, small substitutions of essential amino acids may change binding properties or oligomeric state performance making those enzymes promising candidates for bioengineering.

### 3.2. Monomeric Structure and Dimeric Interface

Members of the OYE family were found to exist in different oligomeric states. A remarkable property observed for class III OYEs is their occurrence in solution as homodimers and homotetramers [22,76,105]. Even higher species as octamers and dodecamers, emerged from functional homodimers, were observed for TOYE and TsOYE [77,79]. A shift between higher multiple oligomeric states was noticed depending on the protein concentration [22,77]. On the other hand, members of class I and class II were found to exclusively occur as monomers or homodimers [14,83,109]. Nevertheless, the basic monomer structure is very similar. A typical single domain is represented by an $(\alpha,\beta)_8$-barrel structure (TIM barrel) with additional secondary structure core elements, in which FMN is bound at the *C*-terminal end (Figure 3). Secondary structure prediction showed that these core elements are similar in location and length in all three classes (Figure S1, SecStruc). An *N*-terminal hairpin of two short β-strands (βA and βB) builds the bottom of the barrel prior to strand β1 (OYE1 numbering) (Figure 3, blue hairpin) [75,78,83].

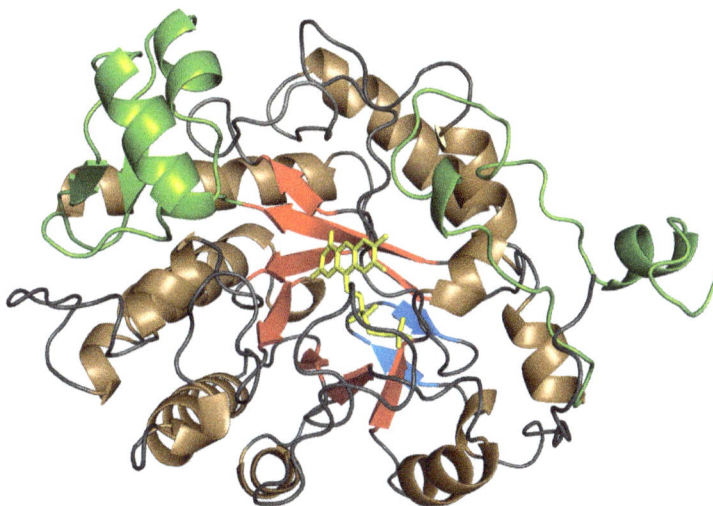

**Figure 3.** TIM-barrel structure of OYEs. Three-dimensional model of the crystal structure of *Saccharomyces pastorianus* OYE1 (pdb entry: 1OYA). The α-helices and β-strands of the TIM-barrel are indicated in gold and red, respectively. The *N*-terminal hairpin of two short β-strands at the bottom of the barrel is indicated in blue. The *C*-terminal helices and the large surface loop between β3 and α3 are indicated in green. The FMN prosthetic group is indicated in yellow.

Additional secondary structure elements occur on surface loops between core building blocks of β-strands and α-helices. The largest loop occurs between β3 and α3 and differs greatly between all OYE representatives (Figure 3, green surface loop) [75,78,83]. Class III enzymes are up to 40 amino acids shorter and therefore more compact. This is due to the shortening of the *N*-terminal surface loops between β3 and α3, but also between β5 and α5 and at the *C*-terminal end of the protein (Figure S1). However, the capping domain between β3 and α3, which is partly covering the entrance of the active site, differs greatly in length and type of structural units within subclass III [78]. Each monomer within the functional dimer is facing the central hole with the same side and contributes the same residues to the hydrogen bond network that keeps monomers together [77]. Residues Gln330 and Tyr331 (TOYE numbering) are highly conserved and seem to play a role in monomer interaction [77]. While Gln330 is specific for class III enzymes, Tyr331 is conserved for all OYEs (Figure S1). A variety of additional residues assist in the formation of the functional dimer–dimer interface such as Thr45, Ser28/His42/Arg46 and Tyr315 (TOYE numbering) [77,78]. Many of them are conserved only for class III enzymes. The most influential factor for the formation of the functional dimer–dimer interface of class III is Arg333 (TOYE numbering). This "arginine finger" stretches into the active site of the adjacent monomer and interacts with the respective flavin cofactor. An additional stabilization of some class III enzymes originating from extremophiles is associated with an increased proline content of surface loops [23,78], and with the presence of three complex salt bridge networks at the dimerization interface that increases the subunit interaction strength. The complex five residue salt bridge between four residues of α2 and an asparagine from α1 is not conserved for class III enzymes. However, incorporation of this salt bridge within another class III enzyme by site-directed mutagenesis indeed increased the thermal stability [22].

*3.3. FMN Binding*

The FMN prosthetic group is bound at the C-terminal end of the β-strands, a typical location for the active sites of TIM-barrel enzymes (Figure 3) [110]. The *re* side of the flavin is facing the protein and is completely hidden from the active site but in contact with strand β1 [111,112]. The *si* side of the flavin is facing a solvent filled access channel (20 Å length in PETNR) and therefore forms the bottom of a wide-open active pocket mainly assembled with aromatic residues [9]. The redox potential of the FMN cofactor is controlled by the different interactions between the protein and the flavin. The N3 and O2 atoms of the flavin pyrimidine ring are interacting with a glutamine (Gln102 in YqjM) and the N1 and O2 with an arginine (Arg215 in YqjM) (Figure 4) [75]. Both residues are strictly conserved in OYEs. Furthermore, a conserved histidine pair is described for all OYEs to be in hydrogen bonding distance. In YqjM (class III), His167 and His164 donate a hydrogen bond for the N1 or N3 atom of the flavin, respectively (Figure 4) [75]. However, in PETNR (class I) the pursuant histidines (His184 and His181) donate the hydrogen bond to the flavin O2 as well as to the activating group of the substrate/inhibitor [111]. While both histidines are highly conserved in class III, the second histidine is replaced by an asparagine in several class I and class II enzymes (Figure 4 and Figure S1). For example, the corresponding asparagine (Asn194) in OYE1 is known to play a key role in ligand binding [83,111]. Moreover, residues Ala60 and Cys26 (YqjM) were found to be in hydrogen bonding distance to the FMN O4 atom (Figure 4) [75]. While Ala60 is conserved within all classes (sometimes glycine in class II), the cysteine residue is a unique feature only for class III enzymes. Cys26 was shown to interact not only with the O4-atom of the flavin but also with Tyr28 (Figure 4), which is also conserved in class III. In class I/II, Cys26 is replaced by a conserved threonine (Thr37 in OYE1), which is as well within hydrogen bonding distance to the O4-atom of the isoalloxazine ring [112]. It was shown that both residues modulate the flavin reduction potential after they were exchanged by mutagenesis [112–114].

The dimethylbenzene moiety of the flavin isoalloxazine ring was described to interact with the hydrophobic residues Met25, Leu311 and Arg308 (YqjM [75]). The methionine is conserved only for class III enzymes as well as the leucine, which is replaced by isoleucine in class I/II. The arginine is also highly conserved for class II but not for class I enzymes. In PETNR, a tyrosine (Tyr351) is in van der Waals contact with the edge of the dimethylbenzene nucleus [111]. This tyrosine is conserved for class I/II. In OYE1 the hydrophobic area around the dimethylbenzene ring is more compact and contains two interacting phenylalanine residues (Phe250 and Phe296). Phe250 is highly conserved in class II enzymes but also occurs in PETNR (Phe240). However, Phe296 is neither conserved in class I nor in class II. Instead of the phenylalanines, class III enzymes possess an arginine (Arg336 in YqjM), a residue involved in both, flavin binding and formation of the dimeric interface (vide supra).

The ribityl chain of the flavin is stabilized in all classes by a conserved proline and an arginine (Pro24 and Arg215: YqjM numbering). Additionally, class III enzymes were also found to anchor the cofactor in this part to Ser23, Ser249 and Gln265 (YqjM numbering) [75]. However, Figure S1 shows that Ser23 is not conserved in class III OYEs, as previously published [9].

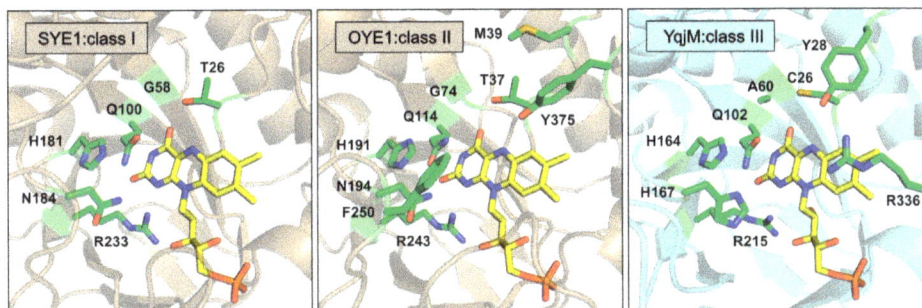

**Figure 4.** Active site of SYE1 (pdb entry: 2QG9; class I) from *Shewanella oneidensis*, OYE1 (pdb entry: 1OYA; class II) from *Saccharomyces pastorianus* and YqjM (pdb entry: 1Z41; class III) from *Bacillus subtilis*. The side chains interacting with the flavin in the active site are shown in stick models and coloured by elements (red = oxygen-containing group; blue = nitrogen containing group). The FMN cofactor is shown as stick model and coloured by elements with carbons in yellow. Class I and class II proteins have very similar active sites and are shown in metallic brown. The class III protein is shown in metallic blue. Note that R336' in YqjM belongs to the adjacent subunit. The rms distances (rmsd) were obtained from overlaid structures (see Figure S2) between SYE1 and OYE1 (0.667 Å), SYE1 and YqjM (1.055 Å), and OYE1 and YqjM (1.262 Å).

*3.4. Coenzyme and Inhibitor Binding*

In co-crystallisation studies of OYEs, the $F_o$–$F_c$ electron density map of the uncomplexed, oxidized enzyme gives a strong positive peak above the flavin isoalloxazine ring. This observation is due to the binding of an anion from the crystallisation solution such as sulphate [75], chloride [83,111], acetate or formate [77]. Addition of NADPH resulted in the replacement of the sulphate by two water molecules and in the reduction of the flavin [75]. In agreement with a bi-bi ping-pong mechanism, great similarity exists among OYEs in the binding mode of the nicotinamide moiety of NADPH and phenolic inhibitors such as *para*-hydroxybenzaldehyde (*p*-HBA) (Figure 5) or *para*-nitrophenol (*p*-NP). Both aromatic rings are oriented through π–π stacking with the FMN isoalloxazine ring and hydrogen bonding with His167/Asn194 and His164/His191 (YqjM/OYE1) [115]. The crystal structure of OYE1 (class II) with an NADP+ analogue showed that the oxygen of the amide on the pyridinium ring is (hydrogen) bonding with the two conserved histidine residues, thus positioning the C4 atom close to the N5 atom of FMN for the hydride transfer [83]. The same position was observed for the nicotinamide ring of tetrahydro-NADH in MR (class I) (Figure 5) and TOYE (class III) [77,116].

Analogues of histidine/asparagine residues are found in all classes of OYE homologues. A remarkable difference is the binding of the functional group of the phenolic ligand. For instance the aldehyde group of *p*-HBA interacts with Tyr375 (OYE1) [83], or Tyr351 (PETNR) [75], respectively. This tyrosine residue is conserved within all OYE classes (Figure S1), but was never mentioned to play a role in the catalysis of class III enzymes, which use a different strategy to build their active sites. A reorientation of the *C*-terminal end causes the formation of a part of the active site of the adjacent monomer. Therefore, only an arginine (Arg333 in TOYE) contributes to the opposed active site. The "arginine finger" causes the formation of a strong hydrogen bond with the nitro-group of *p*-NP [75], or the O1P/O1N atoms of tetrahydro-NADH [77]. Replacement of this arginine might be a biocatalytic tool to broaden/change the substrate spectrum or cofactor specificity of class III enzymes. Another *N*-terminal class III conserved tyrosine (Tyr28 in YqjM) is involved in binding of the aldehyde oxygen of *p*-HBA [75,78]. Furthermore, Tyr175/Tyr196 (TsOYE/OYE1) was confirmed to be a proton donor for the substrate 2-cyclohexenone [75,78], and is conserved within all subclasses. To summarize, the binding partner for the functional group of aromatic substrates is always a tyrosine, which is in hydrogen bonding distance. Class I and class II use two tyrosines from the central part (Tyr196 in

OYE1) and the *C*-terminal protein part (Tyr375 in OYE1), whereas class III enzymes involve the central part tyrosine (Tyr175 in TsOYE) as well, but a second *N*-terminal tyrosine (Tyr25 in TsOYE) takes over the role of the class I/II *C*-terminal tyrosine due to reorientation.

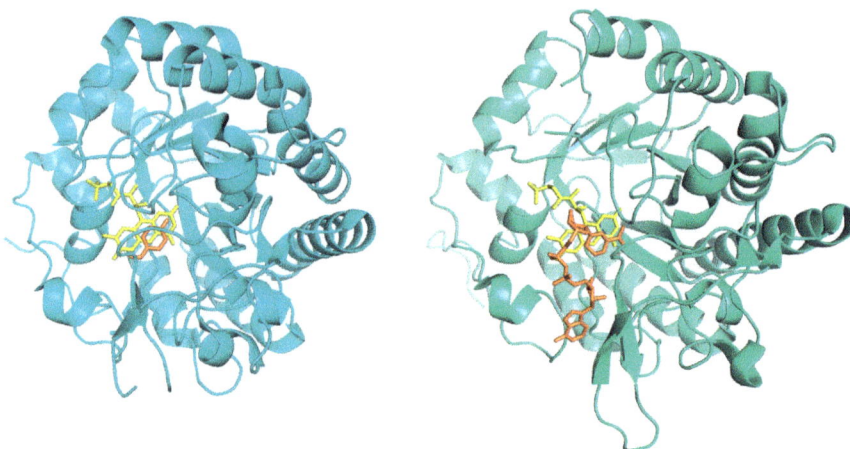

**Figure 5.** Ligand binding in OYEs: (**Left**) NADH:flavin oxidoreductase from *Shewanella oneidensis* (SYE1) with *para*-hydroxybenzaldehyde bound (pdb entry: 2GQ9). The protein is shown in blue cyan, the flavin cofactor in yellow, and the phenolic inhibitor in orange. (**Right**) morphinone reductase (MR) from *Pseudomonas putida* with tetrahydro-NAD bound (pdb entry: 2R14). The protein is shown in green cyan, the flavin cofactor in yellow, and the pyridine nucleotide in orange.

The majority of OYEs display a preference for NADPH over NADH as the source of hydride, as indicated from catalytic efficiencies ($k_{cat}/K_M$) or apparent dissociation constants ($K_D$). The specificity ratio of NADPH:NADH, obtained from specific activity with *trans*-2-hexenal by Bruce and co-workers [117], can vary from 0.02 (MR) to 10 (OYE1), although the cofactor preference for KYE1, XenA and Yers-ER was shown to differ depending on the substrate, yielding different NADPH:NADH specificity ratios, a surprising result given the bi-bi ping-pong mechanism described above [6]. Interestingly, TsOYE displays a similar $K_D$ for NADPH and NADH, but a five times higher reaction rate when using NADPH with respect to NADH [59]. The only OYEs displaying a higher affinity for NADH are from class I: NerA/GTNR [98], MR [85], SYE1 and SYE3 [103]. LacER, which falls short of being assigned to a class, although the closest class it relates to seems to be class I as seen above, also shows a preference for NADH [10].

The group of Hauer recently showed examples of grafted β/α surface loop regions between OYEs MR, NCR and OYE1 that led to altered as well as new reaction activities and a change in NADH interactions [118–120]. One example was showcased with NCR (class I), which displayed lower activity with increasing NAD$^+$ present in the reduction reaction of cinnamaldehyde [118–120]. Through β/α surface loop grafting from OYE1 (class II) and MR (class I) to form various loop variants of NCR, the loss of activity with higher presence of NAD$^+$ was significantly reduced [118–120].

## 4. Reactivity with NCBs

Recently, as mentioned above, NCBs **1–7** in Figure 1 were used as alternative hydride source to replace NAD(P)H [23,52,58,59]. Kinetic data with NCBs (**1–5**) are available for PETNR (class I), and for TOYE, XenA, and TsOYE (class III). Additionally, biocatalytic reactions were performed and conversion data were acquired for class I (yellow): PETNR, LeOPR1, XenB, MR, and NerA; class II (grey): OYE2 and OYE3; and class III (green): XenA, TOYE, TsOYE, DrOYE, and RmOYE (Figure 6) [59]. Other NCBs (**6–7**) were also screened with MR, NCR, OYE1 and OYE3 [58].

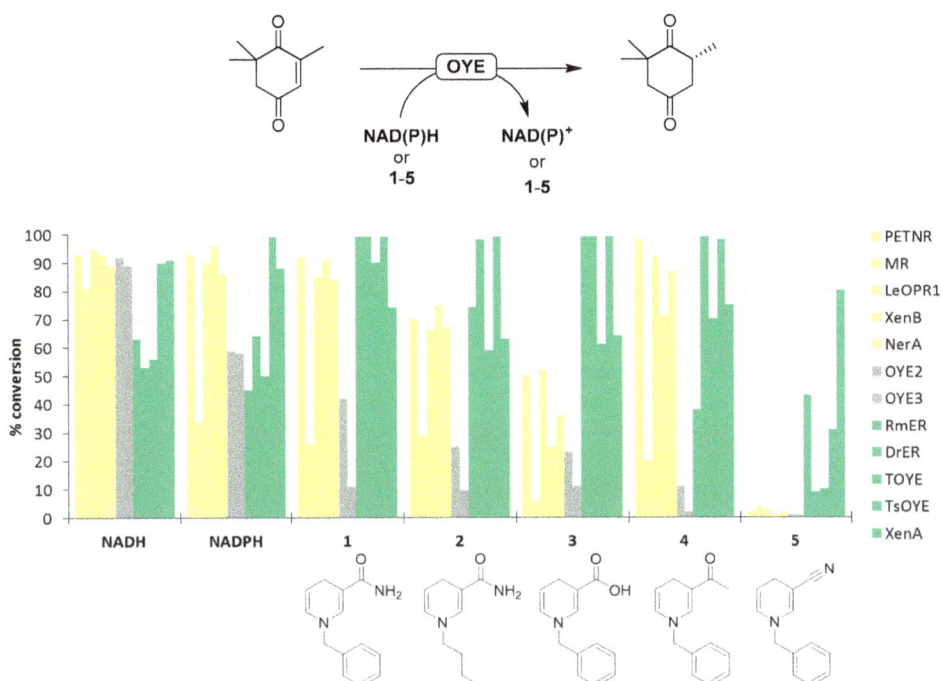

**Figure 6.** Asymmetric hydrogenation of ketoisophorone to (6*R*)-levodione with a panel of OYEs from class I (yellow), II (grey) and III (green) and NCBs **1–5** (1-benzyl-1,4-dihydronicotinamide **1**, 1-butyl-1,4-dihydronicotinamide; **2**, 1-benzyl-1,4-dihydronicotinic acid **3**, 1-benzyl-3-acetyl-1,4-dihydropyridine **4**, and 1-benzyl-3-cyano-1,4-dihydropyridine **5**) (data adapted from [59]).

## 4.1. Biocatalytic Conversions

From the full biocatalytic reaction for the reduction of the model substrate ketoisophorone, moderate to high conversions were obtained with NADPH and NADH and low to high conversions with NCBs **1–5** (Figure 6) [59]. NCB **5** gave very low conversions (1–10%), except for XenA (80%), TsOYE (59%) and RmER (43%) [59]. With our new classification, we noted the enzymes accepting NCB **5** were all from class III, and that OYE2–3 from class II gave lower conversions with the NCBs in general.

## 4.2. Kinetic Data: Steady State and Pre-Steady State

The reduction of 2-cyclohexenone under steady-state conditions gave $k_{cat}$, $K_M$ and catalytic efficiency $k_{cat}/K_M$ for PETNR (class I), TOYE and XenA (class III) [59]. Noting that class III TOYE and XenA gave the highest rates with the NCBs, XenA in particular displayed high catalytic efficiency ($k_{cat}/K_M$) with NCB **2** (Figure 7).

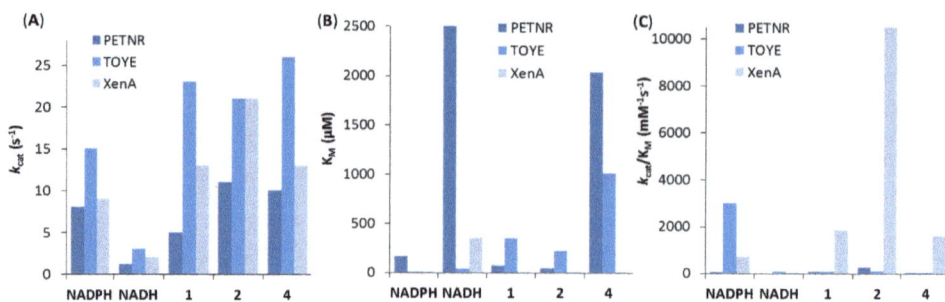

**Figure 7.** Steady-state kinetics with: catalytic activity $k_{cat}$ (**A**); Michaelis constant $K_M$ (**B**); and catalytic efficiency $k_{cat}/K_M$ (**C**) for the reduction of 2-cyclohexenone to 2-cyclohexanone by PETNR, TOYE and XenA with NCBs **1**, **2** and **4** (data adapted from [59]).

The rates of the reductive half-reaction showed that PETNR from class I displayed lower rates when compared to TOYE, XenA and TsOYE from class III (Figure 8A,B) [59]. TOYE gave the highest reduction rates followed by XenA and TsOYE. NCB **3** clearly afforded the highest reduction rate values, but also the highest dissociation constant $K_D$ for the enzyme-reduced nicotinamide complex (Figure 8C,D). For each OYE, the order of the best to poorer NCB differed (Figure 8A,B).

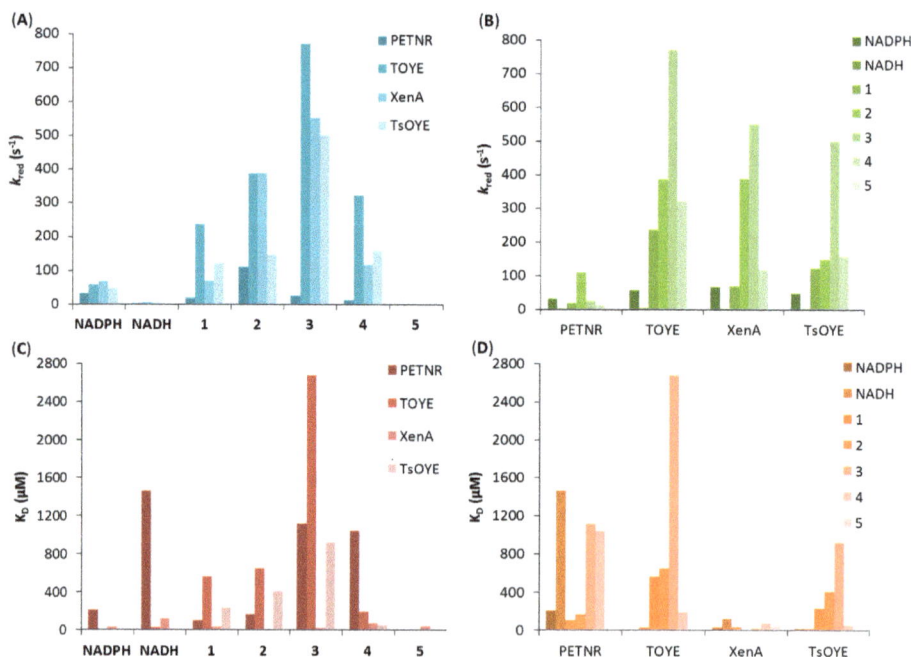

**Figure 8.** Reduction reaction rates $k_{red}$ (**A**,**B**); and dissociation constants $K_D$ (**C**,**D**) for the reductive half-reaction of PETNR, TOYE, XenA and TsOYE and NCBs **1–5** (data adapted from [59]).

Fast reaction kinetics with MR and PETNR (class I) by the group of Scrutton led to the conclusion that quantum mechanical tunnelling is crucial for the hydride transfer from the nicotinamide cofactor to FMN [121]. They proposed that sub-angstrom differences in the donor-acceptor distance affect the probability of hydride transfer [121–123]. From molecular modelling studies with TsOYE and NCBs

1–5 [52], as well as with the crystal structures of XenA [59], the NCBs were observed to be in the correct position, π stacking with the FMN isoalloxazine ring, and the amide oxygen within hydrogen bonding distance of the two conserved histidine residues (His181 and His178 in XenA). The exception was NCB **5**, which contains a nitrile substituent in place of the amide, thus missing the oxygen and only hydrogen bonding with one of the two histidines. This analogue was used as a hydride donor by the OYEs but with less efficiency. Although the coenzyme biomimetics bind to the OYEs' active site in a similar way as NAD(P)H, small changes in the orientation of the nicotinamide ring could affect the rate of hydride transfer. Indeed, different kinetic data was observed for all analogues and OYEs across the three classes. The NCB giving the highest reduction rate for one OYE, NCB **2** for PETNR (class I), was different for another OYE, NCB **3** for TOYE, XenA and TsOYE (class III) [59].

The overlaid crystal structures of XenA (from *P. putida* II-B, class III) with the tetrahydro form of cofactors NADPH$_4$ and NCB **1–2,4** showed only minimal changes. One exception in all three NCBs structures was the alternative conformation of the tryptophan residue (Trp302 in XenA), which reduced the volume of the active site [59].

## 5. Classification of OYEs with Respect to Substrates

As listed in the introduction, OYEs can reduce a variety of α,β-unsaturated activated alkene substrates, which are continuously being extended with new activating groups explored for industrial purposes. In general, the review by Toogood, Gardiner and Scrutton gives an overview of substrate family scopes for OYEs [9]. Looking at the compilation of substrates, there is a large overlap in the substrate scope and stereochemical outcome of OYEs within and across the three classes, and also wide differences within one class. Unfortunately, reported data often varies; in some cases, the specific activity was reported for each substrate, in other cases, percentage conversion of the biocatalytic reactions was reported. Furthermore, reaction times vary (4–48 h) and a range of enzyme concentrations is used, which makes identification of preferred substrates more difficult.

Selecting widely published percentages conversion of biocatalytic reactions on common substrates, we observed significant trends regarding all members of the OYE family but also noticed differences between OYE classes I–III (Figures 9 and 10). OYEs in general convert substituted (methylated) cyclic enones such as 2-methylcyclo-hexenone or -pentenone as well as 3-methylcyclo-hexenone or -pentenone. Comparing 6- and 5-membered ring substrates in Figure 9 shows that ring size is not so important with respect to biocatalytic conversions. It does, however, have a significant effect on the stereochemical outcome of the product (Figure 11). Stereochemical outcome can be influenced by the configuration of the C=C double bond, the orientation of the substrate in the active site and the stereospecificity of the hydrogen addition [124]. For most OYEs, 2-methylcyclopentenone gave the (*S*)-enantiomer whereas 2-methylcyclohexenone afforded the (*R*)-enantiomer (Figure 11). For β-methylated substrates, the conversion was lower in all cases (classes I–III). This effect was repeatedly observed [11,12,42], and is due to steric hindrance. The formed products from β-methylated cyclic enones were (*S*)-enantiomers, independent from the ring size (Figure 11). Interestingly, class I enzymes display significantly lower conversions from the 2-methyl- (80%) to the 3-methyl-cyclohexenone (20%), whereas for class II enzymes this difference goes from 72% to 40%. Class III enzymes have averaged 45% conversion of 2-methylcyclohexenone and none of the tested enzymes were active on the β-methylated substrate. Similar values were found for substituted cyclopentenones. With α-methylated substrates all classes gave a similar average conversion (between 48% and 63%). For β-methyl unsaturated substrates, class III enzymes afforded no conversion, whereas the average conversion of class II enzymes is slightly higher (30%) than that of class I enzymes (20%).

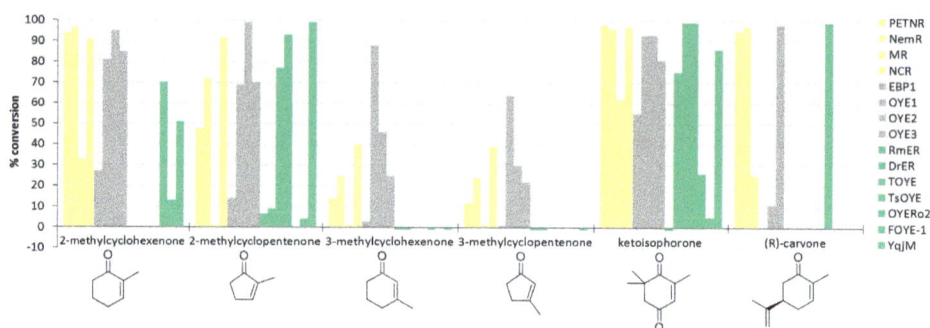

**Figure 9.** Cyclic enones substrate preference of OYEs according to the classification. Negative values represent conversions below 1%. Blank spaces correspond to no measurements. References for conversion values: PETNR [44], NemR [44], MR [44], NCR [42], EBP1 [44], OYE1 [42], OYE2 [42], OYE3 [42], RmER [80], DrER [80], TOYE [77], TsOYE [52], OYERo2 [22], FOYE-1 (unpublished data) and YqjM [12,18].

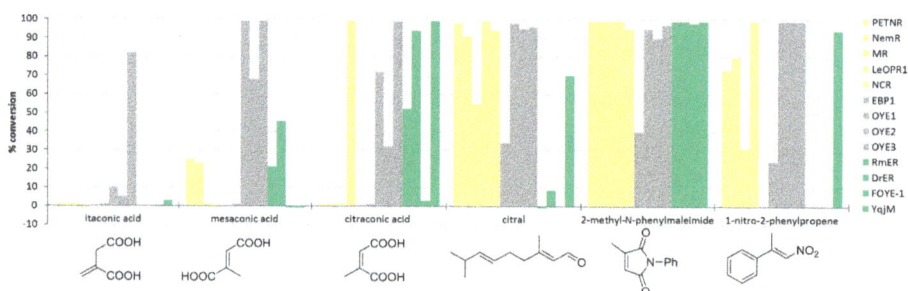

**Figure 10.** Dicarboxylic acids, aldehyde, maleimide, nitroalkenes substrate preference of OYEs according to the classification. Negative values represent conversions below 1%. Blank spaces correspond to no measurements. References for conversion values: PETNR [44], NemR [44], MR [44], LeOPR1 [12,18], NCR [42], EBP1 [44], OYE1 [42], OYE2 [42], OYE3 [42], RmER [80], DrER [80], FOYE-1 (unpublished data) and YqjM [12,18].

Other $\alpha$-substituted unsaturated cyclic ketones such as ketoisophorone and carvones are also alternative substrates for all three classes. However, several non-thermostable class III enzymes such as RmER, OYERo2 [22,80], and FOYE-1 (unpublished data) gave very low conversions on those substrates. Poor conversions are supported by very poor activities towards these typical substrates (ketoisophorone or cyclohexenone), mainly in class III (YqjM, FOYE-1, OYERo2, XenA, TsOYE, DrOYE, and RmOYE) [22,23]. In contrast, these enzymes showed highest specific activities on several maleimides (up to 70 U/mg) [22,23]. Comparably, the specific activities on maleimides for classes I and II enzymes are much lower [6,8,21], but all classes reduce maleimides with high conversion yield (Figure 10). YqiG is highly active towards maleimide (56 U/mg) and shows little activity toward carboxylic acids and cyclic ketones (<3 U/mg) [104]. This non-assigned OYE exhibits high temperature stability and is closely related to class III (see Section 3). With ketoisophorone, (*R*)-carvone, 2-methyl-*N*-phenylmaleimide and 2-methylmaleimide, all investigated OYEs led to the (*R*)-enantiomer (Figure 11). In general the enantiomeric excess was lower with ketoisophorone due to known product racemisation [8]. (Dimethylated) dicarboxylic acids show remarkable differences between the three classes. Itaconic acid was only converted by class II enzymes (average conversion: 25%). The two isomers, mesaconic acid and citraconic acid, were transformed by all three classes.

However, class II showed on average 50% higher conversions with respect to the other two classes and always preferred a distinct enantiomeric product: (*S*)-enantiomer with itaconic and citraconic acid; (*R*)-enantiomer with mesaconic acid. Only with the *cis*-substrate (citraconic acid), class III was the more active enzyme class (Figure 10). Class I enzymes exclusively led to the (*R*)-enantiomer with dimethylated dicarboxylic acids, whereas the stereochemical outcome for class III enzymes was not investigated until now. The aliphatic dimethylated dienal substrate citral was shown to be a great substrate for class I and class II. Only low to modest citral conversions were obtained within class III with YqjM and DrER (Figure 10) [12,80], as well as with TsOYE [52].

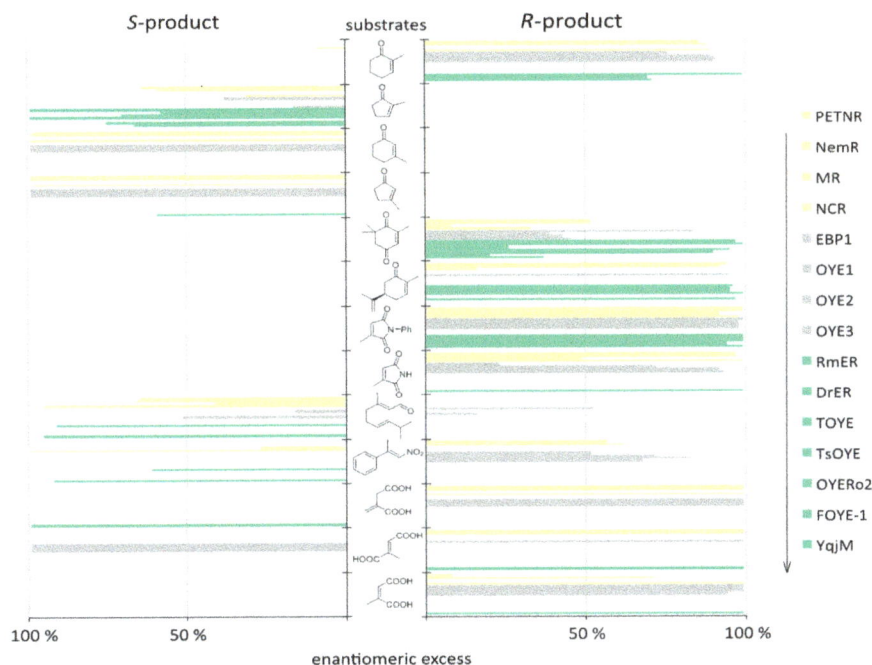

**Figure 11.** Stereochemical outcome of various OYE-catalysed asymmetric hydrogenations. The order of OYEs used for each substrate is from top (PETNR) to bottom (YqjM). Blank spaces correspond to no measurements. References for *ee* values used in this diagram: PETNR [44], NemR [44], MR [44], NCR [42], EBP1 [44], OYE1 [42], OYE2 [42], OYE3 [42], RmER [80], DrER [80], TOYE [77], TsOYE [52], OYERo2 [22], FOYE-1 (unpublished data) and YqjM [12,18].

To summarise, class I and class II prefer cyclic ketones, especially 2-methylcyclohexenone, ketoisophorone and carvones. On dicarboxylic acids, class II enzymes show best conversion rates. Maleimides are excellent substrates for the entire OYE family, class III enzymes especially, displaying remarkable activities on maleimides (about 50-times higher than on all other substrates).

A correlation between the protein structure and the sequence patterns for the stereopreference with respect to reduction of 1-nitro-2-phenylpropene was attempted to be established for 12 independent OYE structures [125]. LeOPR3, NCR, YqjM, XenA, TOYE, TsOYE, and GkOYE reduced 1-nitro-2-phenylpropene to the *S*-enantiomeric product whereas LeOPR1, OYE1, PETNR and NemR produced the (*R*)-enantiomer (Figure 11). However, the group of Pietruszka showed that the stereoselectivity of an uncharacterised OYE could not be predicted solely based on the primary sequence pattern of the surface loop β2 [126].

Stereoselectivity was also investigated by the group of Stewart, who determined how subtle changes in the residues control the orientation of substrate binding in OYEs [127,128], and showed that mutations can lead to inverted enantioselectivity [129,130]. The stereochemical outcome was influenced by mutagenesis of Trp116 in OYE1. For instance, (S)-carvone displayed an inverted stereoselectivity for six different variants. The same observation was made for (Z)-β-aryl-β-cyanoacrylate when Trp116 was changed. Substitution of this amino acid against smaller residues resulted in a classical orientation of the substrate and gave the (S)-enantiomer. Substitution against more bulky residues resulted in a switch of stereoselectivity affording the (R)-enantiomer. This tryptophan is conserved for classes I and II enzymes and might be a major residue for changing enantioselectivity of different prochiral substrates [25]. Class III enzymes possess a conserved alanine at this position. Directed evolution performed on YqjM by Reetz and co-workers enabled control of the enantioselectivity for the bioreduction of substituted cyclopentenone and cyclohexenone substrates [131]. A recent review by Stewart and Amato on protein engineering of OYEs gives a good overview on the possibilities of stereochemical outcome [130].

## 6. Conclusions

We provide a new classification for the currently available and ready to use 63 OYEs, divided into three classes. The structural classification of OYEs displayed the conserved as well as the differing residues among the three classes.

Class I are monomeric OYEs originating from plants and bacteria. They prefer α-methylated cyclic enones as substrates and showed highest reduction rates with NCB **2**. Class II enzymes are fungal OYEs likely developed in a co-driven evolution from class I. Both classes are phylogenetically and structurally closely related resulting in similar substrate preferences, although class II members display higher conversions on certain substrates and those tested did not accept NCBs as well.

Class III OYEs are likely driven by a convergent evolution to class I/II resulting in a substantial distance on sequence level according to the dendrogram. Due to remarkable variances in conservation compared to the two foreign subclasses but great conservation within their own members, class III OYEs have a different structure, biochemical behaviour and substrate preference. They form oligomers built from homodimers and are often thermostable. They highly prefer maleimides as substrates and NCB **3** as the hydride donor, and accept the nitrile-substituted NCB **5** best of all classes.

The symbiosis of bioinformatics, biochemistry and biocatalysis enables a closer analysis of an industrially attractive biocatalyst, OYE, to determine from which class it should be chosen for a targeted product, and to allow further improvements with specially designed NCBs.

**Supplementary Materials:** The following are available online at www.mdpi.com/2073-4344/7/5/130/s1, Figures S1 and S2.

**Acknowledgments:** Caroline E. Paul gratefully acknowledges funding from the NWO Innovative programme Veni (grant number 722.015.011). Anika Scholtissek was financial supported by the Saxon Ministry of Science and Fine Arts and the European Union (EU) in the framework of the European Social Fund (ESF; project numbers 100,101,363 and 100,236,458).

**Author Contributions:** A.S. generated the phylogenetic trees, sequence alignment and analysis; A.H.W. performed the modeling; and A.S., D.T., W.J.H.v.B. and C.E.P. wrote the paper.

**Conflicts of Interest:** The authors declare no conflict of interest. The founding sponsors had no role in the design of the study; in the collection, analyses, or interpretation of data; in the writing of the manuscript, and in the decision to publish the results.

## References

1. Knowles, W.S.; Noyori, R. Pioneering perspectives on asymmetric hydrogenation. *Acc. Chem. Res.* **2007**, *40*, 1238–1239. [CrossRef] [PubMed]
2. Clayden, J.; Greeves, N.; Warren, S.G. *Organic Chemistry*, 2nd ed.; Oxford University Press: Oxford, UK; New York, NY, USA, 2012.

3.  Atkins, P.W.; Shriver, D.F. *Shriver and Atkins' Inorganic Chemistry*, 5th ed.; Oxford University Press: Oxford, UK; New York, NY, USA, 2010.

4.  Yang, J.W.; Fonseca, M.T.H.; Vignola, N.; List, B. Metal-free, organocatalytic asymmetric transfer hydrogenation of α,β-unsaturated aldehydes. *Angew. Chem. Int. Ed.* **2005**, *44*, 108–110. [CrossRef] [PubMed]

5.  Williams, R.E.; Bruce, N.C. 'New uses for an old enzyme'-the old yellow enzyme family of flavoenzymes. *Microbiology* **2002**, *148*, 1607–1614. [CrossRef] [PubMed]

6.  Chaparro-Riggers, J.F.; Rogers, T.A.; Vazquez-Figueroa, E.; Polizzi, K.M.; Bommarius, A.S. Comparison of three enoate reductases and their potential use for biotransformations. *Adv. Synth. Catal.* **2007**, *349*, 1521–1531. [CrossRef]

7.  Stuermer, R.; Hauer, B.; Hall, M.; Faber, K. Asymmetric bioreduction of activated C=C bonds using enoate reductases from the old yellow enzyme family. *Curr. Opin. Chem. Biol.* **2007**, *11*, 203–213. [CrossRef] [PubMed]

8.  Fryszkowska, A.; Toogood, H.; Sakuma, M.; Gardiner, J.M.; Stephens, G.M.; Scrutton, N.S. Asymmetric reduction of activated alkenes by pentaerythritol tetranitrate reductase: Specificity and control of stereochemical outcome by reaction optimisation. *Adv. Synth. Catal.* **2009**, *351*, 2976–2990. [CrossRef] [PubMed]

9.  Toogood, H.S.; Gardiner, J.M.; Scrutton, N.S. Biocatalytic reductions and chemical versatility of the old yellow enzyme family of flavoprotein oxidoreductases. *ChemCatChem* **2010**, *2*, 892–914. [CrossRef]

10. Gao, X.Z.; Ren, J.; Wu, Q.Q.; Zhu, D.M. Biochemical characterization and substrate profiling of a new NADH-dependent enoate reductase from *Lactobacillus casei*. *Enzyme Microb. Technol.* **2012**, *51*, 26–34. [CrossRef] [PubMed]

11. Swiderska, M.A.; Stewart, J.D. Stereoselective enone reductions by *Saccharomyces carlsbergensis* old yellow enzyme. *J. Mol. Catal. B Enzym.* **2006**, *42*, 52–54. [CrossRef]

12. Hall, M.; Stueckler, C.; Ehammer, H.; Pointner, E.; Oberdorfer, G.; Gruber, K.; Hauer, B.; Stuermer, R.; Kroutil, W.; Macheroux, P.; et al. Asymmetric bioreduction of C=C bonds using enoate reductases OPR1, OPR3 and YqjM: Enzyme-based stereocontrol. *Adv. Synth. Catal.* **2008**, *350*, 411–418. [CrossRef]

13. Bertolotti, M.; Brenna, E.; Crotti, M.; Gatti, F.G.; Monti, D.; Parmeggiani, F.; Santangelo, S. Substrate scope evaluation of the enantioselective reduction of β-alkyl-β-arylnitroalkenes by old yellow enzymes 1–3 for organic synthesis applications. *ChemCatChem* **2016**, *8*, 577–583. [CrossRef]

14. Nivinskas, H.; Sarlauskas, J.; Anusevicius, Z.; Toogood, H.S.; Scrutton, N.S.; Cenas, N. Reduction of aliphatic nitroesters and *N*-nitramines by *Enterobacter cloacae* PB2 pentaerythritol tetranitrate reductase. *FEBS J.* **2008**, *275*, 6192–6203. [CrossRef] [PubMed]

15. Toogood, H.S.; Fryszkowska, A.; Hare, V.; Fisher, K.; Roujeinikova, A.; Leys, D.; Gardiner, J.M.; Stephens, G.M.; Scrutton, N.S. Structure-based insight into the asymmetric bioreduction of the C=C double bond of α,β-unsaturated nitroalkenes by pentaerythritol tetranitrate reductase. *Adv. Synth. Catal.* **2008**, *350*, 2789–2803. [CrossRef] [PubMed]

16. Liu, Y.J.; Pei, X.Q.; Lin, H.; Gai, P.; Liu, Y.C.; Wu, Z.L. Asymmetric bioreduction of activated alkenes by a novel isolate of *Achromobacter* species producing enoate reductase. *Appl. Microbiol. Biotechnol.* **2012**, *95*, 635–645. [CrossRef] [PubMed]

17. Wang, H.B.; Pei, X.Q.; Wu, Z.L. An enoate reductase *Achr*-OYE4 from *Achromobacter* sp. JA81: Characterization and application in asymmetric bioreduction of C=C bonds. *Appl. Microbiol. Biotechnol.* **2014**, *98*, 705–715. [CrossRef] [PubMed]

18. Stueckler, C.; Hall, M.; Ehammer, H.; Pointner, E.; Kroutil, W.; Macheroux, P.; Faber, K. Stereocomplementary bioreduction of α,β-unsaturated dicarboxylic acids and dimethyl esters using enoate reductases: Enzyme-and substrate-based stereocontrol. *Org. Lett.* **2007**, *9*, 5409–5411. [CrossRef] [PubMed]

19. Tasnádi, G.; Winkler, C.K.; Clay, D.; Sultana, N.; Fabian, W.M.F.; Hall, M.; Ditrich, K.; Faber, K. A substrate-driven approach to determine reactivities of α,β-unsaturated carboxylic esters towards asymmetric bioreduction. *Chem. Eur. J.* **2012**, *18*, 10362–10367. [CrossRef] [PubMed]

20. Tasnádi, G.; Winkler, C.K.; Clay, D.; Hall, M.; Faber, K. Reductive dehalogenation of β-haloacrylic ester derivatives mediated by ene-reductases. *Catal. Sci. Technol.* **2012**, *2*, 1548–1552. [CrossRef]

21. Fu, Y.L.; Hoelsch, K.; Weuster-Botz, D. A novel ene-reductase from *Synechococcus* sp. PCC 7942 for the asymmetric reduction of alkenes. *Process. Biochem.* **2012**, *47*, 1988–1997. [CrossRef]

22. Riedel, A.; Mehnert, M.; Paul, C.E.; Westphal, A.H.; van Berkel, W.J.H.; Tischler, D. Functional characterization and stability improvement of a 'thermophilic-like' ene-reductase from *Rhodococcus opacus* 1CP. *Front. Microbiol.* **2015**, *6*, 1073. [CrossRef] [PubMed]

23. Scholtissek, A.; Ullrich, S.R.; Mühling, M.; Schlömann, M.; Paul, C.E.; Tischler, D. A thermophilic-like ene-reductase originating from an acidophilic iron oxidizer. *Appl. Microbiol. Biotechnol.* **2017**, *101*, 609–619. [CrossRef] [PubMed]

24. Winkler, C.K.; Clay, D.; Turrini, N.G.; Lechner, H.; Kroutil, W.; Davies, S.; Debarge, S.; O'Neill, P.; Steflik, J.; Karmilowicz, M.; et al. Nitrile as activating group in the asymmetric bioreduction of β-cyanoacrylic acids catalyzed by ene-reductases. *Adv. Synth. Catal.* **2014**, *356*, 1878–1882. [CrossRef] [PubMed]

25. Brenna, E.; Crotti, M.; Gatti, F.G.; Monti, D.; Parmeggiani, F.; Powell, R.W.; Santangelo, S.; Stewart, J.D. Opposite enantioselectivity in the bioreduction of (Z)-β-aryl-β-cyanoacrylates mediated by the tryptophan 116 mutants of old yellow enzyme 1: Synthetic approach to (R)-and (S)-β-aryl-γ-lactams. *Adv. Synth. Catal.* **2015**, *357*, 1849–1860. [CrossRef]

26. Swiderska, M.A.; Stewart, J.D. Asymmetric bioreductions of β-nitro acrylates as a route to chiral $\beta^2$-amino acids. *Org. Lett.* **2006**, *8*, 6131–6133. [CrossRef] [PubMed]

27. Durchschein, K.; Hall, M.; Faber, K. Unusual reactions mediated by FMN-dependent ene- and nitro-reductases. *Green Chem.* **2013**, *15*, 1764–1772. [CrossRef]

28. Wohlgemuth, R. Biocatalysis-key to sustainable industrial chemistry. *Curr. Opin. Biotechnol.* **2010**, *21*, 713–724. [CrossRef] [PubMed]

29. Bougioukou, D.J.; Walton, A.Z.; Stewart, J.D. Towards preparative-scale, biocatalytic alkene reductions. *Chem. Commun.* **2010**, *46*, 8558–8560. [CrossRef] [PubMed]

30. Huisman, G.W.; Collier, S.J. On the development of new biocatalytic processes for practical pharmaceutical synthesis. *Curr. Opin. Chem. Biol.* **2013**, *17*, 284–292. [CrossRef] [PubMed]

31. Ress, T.; Hummel, W.; Hanlon, S.P.; Iding, H.; Gröger, H. The organic-synthetic potential of recombinant ene reductases: Substrate-scope evaluation and process optimization. *ChemCatChem* **2015**, *7*, 1302–1311. [CrossRef]

32. Pietruszka, J.; Scholzel, M. Ene reductase-catalysed synthesis of (R)-profen derivatives. *Adv. Synth. Catal.* **2012**, *354*, 751–756. [CrossRef]

33. Winkler, C.K.; Clay, D.; Davies, S.; O'Neill, P.; McDaid, P.; Debarge, S.; Steflik, J.; Karmilowicz, M.; Wong, J.W.; Faber, K. Chemoenzymatic asymmetric synthesis of pregabalin precursors via asymmetric bioreduction of β-cyanoacrylate esters using ene-reductases. *J. Org. Chem.* **2013**, *78*, 1525–1533. [CrossRef] [PubMed]

34. Turrini, N.G.; Hall, M.; Faber, K. Enzymatic synthesis of optically active lactones via asymmetric bioreduction using ene-reductases from the old yellow enzyme family. *Adv. Synth. Catal.* **2015**, *357*, 1861–1871. [CrossRef]

35. Collins, I. Saturated and unsaturated lactones. *J. Chem. Soc. Perkin Trans.* **1999**, *1*, 1377–1395. [CrossRef]

36. Toogood, H.S.; Scrutton, N.S. New developments in 'ene'-reductase catalysed biological hydrogenations. *Curr. Opin. Chem. Biol.* **2014**, *19*, 107–115. [CrossRef] [PubMed]

37. Breithaupt, C.; Strassner, J.; Breitinger, U.; Huber, R.; Macheroux, P.; Schaller, A.; Clausen, T. X-ray structure of 12-oxophytodienoate reductase 1 provides structural insight into substrate binding and specificity within the family of OYE. *Structure* **2001**, *9*, 419–429. [CrossRef]

38. Kohli, R.M.; Massey, V. The oxidative half-reaction of old yellow enzyme-The role of tyrosine 196. *J. Biol. Chem.* **1998**, *273*, 32763–32770. [CrossRef] [PubMed]

39. Lonsdale, R.; Reetz, M.T. Reduction of α,β-unsaturated ketones by old yellow enzymes: Mechanistic insights from quantum mechanics/molecular mechanics calculations. *J. Am. Chem. Soc.* **2015**, *137*, 14733–14742. [CrossRef] [PubMed]

40. Karplus, P.A.; Fox, K.M.; Massey, V. Flavoprotein structure and mechanism. 8. Structure-function relations for old yellow enzyme. *FASEB J.* **1995**, *9*, 1518–1526. [PubMed]

41. Brown, B.J.; Hyun, J.W.; Duvvuri, S.; Karplus, P.A.; Massey, V. The role of glutamine 114 in old yellow enzyme. *J. Biol. Chem.* **2002**, *277*, 2138–2145. [CrossRef] [PubMed]

42. Hall, M.; Stueckler, C.; Hauer, B.; Stuermer, R.; Friedrich, T.; Breuer, M.; Kroutil, W.; Faber, K. Asymmetric bioreduction of activated C=C bonds using *Zymomonas mobilis* NCR enoate reductase and old yellow enzymes OYE 1–3 from yeasts. *Eur. J. Org. Chem.* **2008**, *1511–1516*. [CrossRef]

43. Hall, M.; Stueckler, C.; Kroutil, W.; Macheroux, P.; Faber, K. Asymmetric bioreduction of activated alkenes using cloned 12-oxophytodienoate reductase isoenzymes OPR-1 and OPR-3 from *Lycopersicon esculentum* (tomato): A striking change of stereoselectivity. *Angew. Chem. Int. Ed.* **2007**, *46*, 3934–3937. [CrossRef] [PubMed]

44. Mueller, N.J.; Stueckler, C.; Hauer, B.; Baudendistel, N.; Housden, H.; Bruce, N.C.; Faber, K. The substrate spectra of pentaerythritol tetranitrate reductase, morphinone reductase, *N*-ethylmaleimide reductase and estrogen-binding protein in the asymmetric bioreduction of activated alkenes. *Adv. Synth. Catal.* **2010**, *352*, 387–394. [CrossRef]

45. Tauber, K.; Hall, M.; Kroutil, W.; Fabian, W.M.F.; Faber, K.; Glueck, S.M. A highly efficient ADH-coupled NADH-recycling system for the asymmetric bioreduction of carbon-carbon double bonds using enoate reductases. *Biotechnol. Bioeng.* **2011**, *108*, 1462–1467. [CrossRef] [PubMed]

46. Toogood, H.S.; Knaus, T.; Scrutton, N.S. Alternative hydride sources for ene-reductases: Current trends. *ChemCatChem* **2014**, *6*, 951–954. [CrossRef]

47. Peers, M.K.; Toogood, H.S.; Heyes, D.J.; Mansell, D.; Coe, B.J.; Scrutton, N.S. Light-driven biocatalytic reduction of α,β-unsaturated compounds by ene reductases employing transition metal complexes as photosensitizers. *Catal. Sci. Technol.* **2016**, *6*, 169–177. [CrossRef] [PubMed]

48. Winkler, C.K.; Clay, D.; Entner, M.; Plank, M.; Faber, K. NAD(P)H-independent asymmetric C=C bond reduction catalyzed by ene reductases by using artificial co-substrates as the hydrogen donor. *Chem. Eur. J.* **2014**, *20*, 1403–1409. [CrossRef] [PubMed]

49. Winkler, C.K.; Clay, D.; van Heerden, E.; Faber, K. Overcoming co-product inhibition in the nicotinamide independent asymmetric bioreduction of activated C=C-bonds using flavin-dependent ene-reductases. *Biotechnol. Bioeng.* **2013**, *110*, 3085–3092. [CrossRef] [PubMed]

50. Durchschein, K.; Wallner, S.; Macheroux, P.; Schwab, W.; Winkler, T.; Kreis, W.; Faber, K. Nicotinamide-dependent ene reductases as alternative biocatalysts for the reduction of activated alkenes. *Eur. J. Org. Chem.* **2012**, 4963–4968. [CrossRef]

51. Stueckler, C.; Reiter, T.C.; Baudendistel, N.; Faber, K. Nicotinamide-independent asymmetric bioreduction of C=C-bonds via disproportionation of enones catalyzed by enoate reductases. *Tetrahedron* **2010**, *66*, 663–667. [CrossRef] [PubMed]

52. Paul, C.E.; Gargiulo, S.; Opperman, D.J.; Lavandera, I.; Gotor-Fernández, V.; Gotor, V.; Taglieber, A.; Arends, I.W.C.E.; Hollmann, F. Mimicking nature: Synthetic nicotinamide cofactors for C=C bioreduction using enoate reductases. *Org. Lett.* **2013**, *15*, 180–183. [CrossRef] [PubMed]

53. Paul, C.E.; Arends, I.W.C.E.; Hollmann, F. Is simpler better? Synthetic nicotinamide cofactor analogues for redox chemistry. *ACS Catal.* **2014**, *4*, 788–797. [CrossRef]

54. Hollmann, F.; Paul, C.E. Synthetische nikotinamide in der biokatalyse. *BIOspektrum* **2015**, *21*, 376–378. [CrossRef]

55. Paul, C.E.; Hollmann, F. A survey of synthetic nicotinamide cofactors in enzymatic processes. *Appl. Microbiol. Biotechnol.* **2016**, *100*, 4773–4778. [CrossRef] [PubMed]

56. Okamoto, Y.; Köhler, V.; Paul, C.E.; Hollmann, F.; Ward, T.R. Efficient in situ regeneration of NADH mimics by an artificial metalloenzyme. *ACS Catal.* **2016**, *6*, 3553–3557. [CrossRef]

57. Geddes, A.; Paul, C.E.; Hay, S.; Hollmann, F.; Scrutton, N.S. Donor-acceptor distance sampling enhances the performance of "better than Nature" nicotinamide coenzyme biomimetics. *J. Am. Chem. Soc.* **2016**, *138*, 11089–11092. [CrossRef] [PubMed]

58. Löw, S.A.; Löw, I.M.; Weissenborn, M.J.; Hauer, B. Enhanced ene-reductase activity through alteration of artificial nicotinamide cofactor substituents. *ChemCatChem* **2016**, *8*, 911–915. [CrossRef]

59. Knaus, T.; Paul, C.E.; Levy, C.W.; de Vries, S.; Mutti, F.G.; Hollmann, F.; Scrutton, N.S. Better than Nature: Nicotinamide biomimetics that outperform natural coenzymes. *J. Am. Chem. Soc.* **2016**, *138*, 1033–1039. [CrossRef] [PubMed]

60. Winkler, C.K.; Tasnádi, G.; Clay, D.; Hall, M.; Faber, K. Asymmetric bioreduction of activated alkenes to industrially relevant optically active compounds. *J. Biotechnol.* **2012**, *162*, 381–389. [CrossRef] [PubMed]

61. Saito, K.; Thiele, D.J.; Davio, M.; Lockridge, O.; Massey, V. The cloning and expression of a gene encoding old yellow enzyme from *Saccharomyces carlsbergensis*. *J. Biol. Chem.* **1991**, *266*, 20720–20724. [PubMed]

62. Kataoka, M.; Kotaka, A.; Hasegawa, A.; Wada, M.; Yoshizumi, A.; Nakamori, S.; Shimizu, S. Old yellow enzyme from *Candida macedoniensis* catalyzes the stereospecific reduction of the C=C bond of ketoisophorone. *Biosci. Biotechnol. Biochem.* **2002**, *66*, 2651–2657. [CrossRef] [PubMed]

63. Nizam, S.; Verma, S.; Borah, N.N.; Gazara, R.K.; Verma, P.K. Comprehensive genome-wide analysis reveals different classes of enigmatic old yellow enzyme in fungi. *Sci. Rep.* **2014**, *4*, 4013. [CrossRef] [PubMed]

64. Nizam, S.; Gazara, R.K.; Verma, S.; Singh, K.; Verma, P.K. Comparative structural modeling of six old yellow enzymes (OYEs) from the necrotrophic fungus *Ascochyta rabiei*: Insight into novel OYE classes with differences in cofactor binding, organization of active site residues and stereopreferences. *PLoS ONE* **2014**, *9*, e95989. [CrossRef] [PubMed]

65. Schaller, F.; Weiler, E.W. Molecular cloning and characterization of 12-oxophytodienoate reductase, an enzyme of the octadecanoid signaling pathway from *Arabidopsis thaliana*-Structural and functional relationship to yeast old yellow enzyme. *J. Biol. Chem.* **1997**, *272*, 28066–28072. [CrossRef] [PubMed]

66. Müssig, C.; Biesgen, C.; Lisso, J.; Uwer, U.; Weiler, E.W.; Altmann, T. A novel stress-inducible 12-oxophytodienoate reductase from *Arabidopsis thaliana* provides a potential link between Brassinosteroid-action and Jasmonic-acid synthesis. *J. Plant Physiol.* **2000**, *157*, 143–152. [CrossRef]

67. Strassner, J.; Schaller, F.; Frick, U.B.; Howe, G.A.; Weiler, E.W.; Amrhein, N.; Macheroux, P.; Schaller, A. Characterization and cDNA-microarray expression analysis of 12-oxophytodienoate reductases reveals differential roles for octadecanoid biosynthesis in the local versus the systemic wound response. *Plant J.* **2002**, *32*, 585–601. [CrossRef] [PubMed]

68. Strassner, J.; Furholz, A.; Macheroux, P.; Amrhein, N.; Schaller, A. A homolog of old yellow enzyme in tomato-Spectral properties and substrate specificity of the recombinant protein. *J. Biol. Chem.* **1999**, *274*, 35067–35073. [CrossRef] [PubMed]

69. French, C.E.; Nicklin, S.; Bruce, N.C. Sequence and properties of pentaerythritol tetranitrate reductase from *Enterobacter cloacae* PB2. *J. Bacteriol.* **1996**, *178*, 6623–6627. [CrossRef] [PubMed]

70. Blehert, D.S.; Knoke, K.L.; Fox, B.G.; Chambliss, G.H. Regioselectivity of nitroglycerin denitration by flavoprotein nitroester reductases purified from two *Pseudomonas* species. *J. Bacteriol.* **1997**, *179*, 6912–6920. [CrossRef] [PubMed]

71. Husserl, J.; Hughes, J.B.; Spain, J.C. Key enzymes enabling the growth of *Arthrobacter* sp. strain JBH1 with nitroglycerin as the sole source of carbon and nitrogen. *Appl. Environ. Microbiol.* **2012**, *78*, 3649–3655. [CrossRef] [PubMed]

72. Khairy, H.; Wübbeler, J.H.; Steinbüchel, A. The NADH: Flavin oxidoreductase Nox from *Rhodococcus erythropolis* MI2 is the key enzyme of 4,4′-dithiodibutyric acid degradation. *Lett. Appl. Microbiol.* **2016**, *63*, 434–441. [CrossRef] [PubMed]

73. Pei, X.Q.; Xu, M.Y.; Wu, Z.L. Two "classical" old yellow enzymes from *Chryseobacterium* sp. CA49: Broad substrate specificity of *Chr*-OYE1 and limited activity of *Chr*-OYE2. *J. Mol. Catal. B Enzym.* **2016**, *123*, 91–99. [CrossRef]

74. Xu, M.Y.; Pei, X.Q.; Wu, Z.L. Identification and characterization of a novel "thermophilic-like" old yellow enzyme from the genome of *Chryseobacterium* sp. CA49. *J. Mol. Catal. B Enzym.* **2014**, *108*, 64–71. [CrossRef]

75. Kitzing, K.; Fitzpatrick, T.B.; Wilken, C.; Sawa, J.; Bourenkov, G.P.; Macheroux, P.; Clausen, T. The 1.3 Å crystal structure of the flavoprotein YqjM reveals a novel class of old yellow enzymes. *J. Biol. Chem.* **2005**, *280*, 27904–27913. [CrossRef] [PubMed]

76. Schittmayer, M.; Glieder, A.; Uhl, M.K.; Winkler, A.; Zach, S.; Schrittwieser, J.H.; Kroutil, W.; Macheroux, P.; Gruber, K.; Kambourakis, S.; et al. Old yellow enzyme-catalyzed dehydrogenation of saturated ketones. *Adv. Synth. Catal.* **2011**, *353*, 268–274. [CrossRef]

77. Adalbjörnsson, B.V.; Toogood, H.S.; Fryszkowska, A.; Pudney, C.R.; Jowitt, T.A.; Leys, D.; Scrutton, N.S. Biocatalysis with thermostable enzymes: Structure and properties of a thermophilic 'ene'-reductase related to old yellow enzyme. *ChemBioChem* **2010**, *11*, 197–207. [CrossRef] [PubMed]

78. Opperman, D.J.; Sewell, B.T.; Litthauer, D.; Isupov, M.N.; Littlechild, J.A.; van Heerden, E. Crystal structure of a thermostable old yellow enzyme from *Thermus scotoductus* SA-01. *Biochem. Biophys. Res. Commun.* **2010**, *393*, 426–431. [CrossRef] [PubMed]

79. Opperman, D.J.; Piater, L.A.; van Heerden, E. A novel chromate reductase from *Thermus scotoductus* SA-01 related to old yellow enzyme. *J. Bacteriol.* **2008**, *190*, 3076–3082. [CrossRef] [PubMed]

80. Litthauer, S.; Gargiulo, S.; van Heerden, E.; Hollmann, F.; Opperman, D.J. Heterologous expression and characterization of the ene-reductases from *Deinococcus radiodurans* and *Ralstonia metallidurans*. *J. Mol. Catal. B Enzym.* **2014**, *99*, 89–95. [CrossRef]

81. Fu, Y.L.; Castiglione, K.; Weuster-Botz, D. Comparative characterization of novel ene-reductases from cyanobacteria. *Biotechnol. Bioeng.* **2013**, *110*, 1293–1301. [CrossRef] [PubMed]

82. Warburg, O.; Christian, W. Yellow enzyme and its effects. *Biochem. Z.* **1933**, *266*, 377–411.

83. Fox, K.M.; Karplus, P.A. Old yellow enzyme at 2 Å resolution: Overall structure, ligand-binding, and comparison with related flavoproteins. *Structure* **1994**, *2*, 1089–1105. [CrossRef]

84. Müller, A.; Hauer, B.; Rosche, B. Asymmetric alkene reduction by yeast old yellow enzymes and by a novel *Zymomonas mobilis* reductase. *Biotechnol. Bioeng.* **2007**, *98*, 22–29. [CrossRef] [PubMed]

85. French, C.E.; Bruce, N.C. Purification and characterization of morphinone reductase from *Pseudomonas putida* M10. *Biochem. J.* **1994**, *301*, 97–103. [CrossRef] [PubMed]

86. Haas, E. Isolation of a new yellow enzyme. *Biochem. Z.* **1938**, *298*, 378–390.

87. Stott, K.; Saito, K.; Thiele, D.J.; Massey, V. Old yellow enzyme-the discovery of multiple isozymes and a family of related proteins. *J. Biol. Chem.* **1993**, *268*, 6097–6106. [PubMed]

88. Niino, Y.S.; Chakraborty, S.; Brown, B.J.; Massey, V. A new old yellow enzyme of *Saccharomyces cerevisiae*. *J. Biol. Chem.* **1995**, *270*, 1983–1991. [PubMed]

89. Madani, N.D.; Malloy, P.J.; Rodriguezpombo, P.; Krishnan, A.V.; Feldman, D. *Candida albicans* estrogen-binding protein gene encodes an oxidoreductase that is inhibited by estradiol. *Proc. Natl. Acad. Sci. USA* **1994**, *91*, 922–926. [CrossRef] [PubMed]

90. Komduur, J.A.; Leao, A.N.; Monastyrska, I.; Veenhuis, M.; Kiel, J.A.K.W. Old yellow enzyme confers resistance of *Hansenula polymorpha* towards allyl alcohol. *Curr. Genet.* **2002**, *41*, 401–406. [CrossRef] [PubMed]

91. Kataoka, M.; Kotaka, A.; Thiwthong, R.; Wada, M.; Nakamori, S.; Shimizu, S. Cloning and overexpression of the old yellow enzyme gene of *Candida macedoniensis*, and its application to the production of a chiral compound. *J. Biotechnol.* **2004**, *114*, 1–9. [CrossRef] [PubMed]

92. Miranda, M.; Ramirez, J.; Guevara, S.; Ongaylarios, L.; Pena, A.; Coria, R. Nucleotide-sequence and chromosomal localization of the gene encoding the old yellow enzyme from *Kluyveromyces lactis*. *Yeast* **1995**, *11*, 459–465. [CrossRef] [PubMed]

93. Padhi, S.K.; Bougioukou, D.J.; Stewart, J.D. Site-saturation mutagenesis of tryptophan 116 of *Saccharomyces pastorianus* old yellow enzyme uncovers stereocomplementary variants. *J. Am. Chem. Soc.* **2009**, *131*, 3271–3280. [CrossRef] [PubMed]

94. Patterson-Orazem, A.; Sullivan, B.; Stewart, J.D. *Pichia stipitis* OYE 2.6 variants with improved catalytic efficiencies from site-saturation mutagenesis libraries. *Bioorg. Med. Chem.* **2014**, *22*, 5628–5632. [CrossRef] [PubMed]

95. Ni, Y.; Yu, H.L.; Lin, G.Q.; Xu, J.H. An ene reductase from *Clavispora lusitaniae* for asymmetric reduction of activated alkenes. *Enzyme Microb. Technol.* **2014**, *56*, 40–45. [CrossRef] [PubMed]

96. Zhang, B.Q.; Zheng, L.D.; Lin, J.P.; Wei, D.Z. Characterization of an ene-reductase from *Meyerozyma guilliermondii* for asymmetric bioreduction of α,β-unsaturated compounds. *Biotechnol. Lett.* **2016**, *38*, 1527–1534. [CrossRef] [PubMed]

97. Biesgen, C.; Weiler, E.W. Structure and regulation of OPR1 and OPR2, two closely related genes encoding 12-oxophytodienoic acid-10,11-reductases from *Arabidopsis thaliana*. *Planta* **1999**, *208*, 155–165. [CrossRef] [PubMed]

98. Snape, J.R.; Walkley, N.A.; Morby, A.P.; Nicklin, S.; White, G.F. Purification, properties, and sequence of glycerol trinitrate reductase from *Agrobacterium radiobacter*. *J. Bacteriol.* **1997**, *179*, 7796–7802. [CrossRef] [PubMed]

99. Richter, N.; Gröger, H.; Hummel, W. Asymmetric reduction of activated alkenes using an enoate reductase from *Gluconobacter oxydans*. *Appl. Microbiol. Biotechnol.* **2011**, *89*, 79–89. [CrossRef] [PubMed]

100. Miura, K.; Tomioka, Y.; Suzuki, H.; Yonezawa, M.; Hishinuma, T.; Mizugaki, M. Molecular cloning of the nemA gene encoding N-ethylmaleimide reductase from *Escherichia coli*. *Biol. Pharm. Bull.* **1997**, *20*, 110–112. [CrossRef] [PubMed]

101. Peters, C.; Kolzsch, R.; Kadow, M.; Skalden, L.; Rudroff, F.; Mihovilovic, M.D.; Bornscheuer, U.T. Identification, characterization, and application of three enoate reductases from *Pseudomonas putida* in in vitro enzyme cascade reactions. *ChemCatChem* **2014**, *6*, 1021–1027. [CrossRef]

102. Blehert, D.S.; Fox, B.G.; Chambliss, G.H. Cloning and sequence analysis of two *Pseudomonas* flavoprotein xenobiotic reductases. *J. Bacteriol.* **1999**, *181*, 6254–6263. [PubMed]
103. Brigé, A.; van den Hemel, D.; Carpentier, W.; De Smet, L.; Van Beeumen, J.J. Comparative characterization and expression analysis of the four old yellow enzyme homologues from *Shewanella oneidensis* indicate differences in physiological function. *Biochem. J.* **2006**, *394*, 335–344. [CrossRef] [PubMed]
104. Sheng, X.Q.; Yan, M.; Xu, L.; Wei, M. Identification and characterization of a novel old yellow enzyme from *Bacillus subtilis* str.168. *J. Mol. Catal. B Enzym.* **2016**, *130*, 18–24. [CrossRef]
105. Tsuji, N.; Honda, K.; Wada, M.; Okano, K.; Ohtake, H. Isolation and characterization of a thermotolerant ene reductase from *Geobacillus* sp. 30 and its heterologous expression in *Rhodococcus opacus*. *Appl. Microbiol. Biotechnol.* **2014**, *98*, 5925–5935. [CrossRef] [PubMed]
106. Fitzpatrick, T.B.; Amrhein, N.; Macheroux, P. Characterization of YqjM, an old yellow enzyme homolog from *Bacillus subtilis* involved in the oxidative stress response. *J. Biol. Chem.* **2003**, *278*, 19891–19897. [CrossRef] [PubMed]
107. Brooks, D.J.; Fresco, J.R.; Lesk, A.M.; Singh, M. Evolution of amino acid frequencies in proteins over deep time: Inferred order of introduction of amino acids into the genetic code. *Mol. Biol. Evol.* **2002**, *19*, 1645–1655. [CrossRef] [PubMed]
108. Longo, L.M.; Blaber, M. Protein design at the interface of the pre-biotic and biotic worlds. *Arch. Biochem. Biophys.* **2012**, *526*, 16–21. [CrossRef] [PubMed]
109. Barna, T.; Messiha, H.L.; Petosa, C.; Bruce, N.C.; Scrutton, N.S.; Moody, P.C.E. Crystal structure of bacterial morphinone reductase and properties of the C191A mutant enzyme. *J. Biol. Chem.* **2002**, *277*, 30976–30983. [CrossRef] [PubMed]
110. Wierenga, R.K. The TIM-barrel fold: A versatile framework for efficient enzymes. *FEBS Lett.* **2001**, *492*, 193–198. [CrossRef]
111. Barna, T.M.; Khan, H.; Bruce, N.C.; Barsukov, I.; Scrutton, N.S.; Moody, P.C.E. Crystal structure of pentaerythritol tetranitrate reductase: "Flipped" binding geometries for steroid substrates in different redox states of the enzyme. *J. Mol. Biol.* **2001**, *310*, 433–447. [CrossRef] [PubMed]
112. Spiegelhauer, O.; Werther, T.; Mende, S.; Knauer, S.H.; Dobbek, H. Determinants of substrate binding and protonation in the flavoenzyme Xenobiotic reductase A. *J. Mol. Biol.* **2010**, *403*, 286–298. [CrossRef] [PubMed]
113. Xu, D.; Kohli, R.M.; Massey, V. The role of threonine 37 in flavin reactivity of the old yellow enzyme. *Proc. Natl. Acad. Sci. USA* **1999**, *96*, 3556–3561. [CrossRef] [PubMed]
114. Messiha, H.L.; Bruce, N.C.; Sattelle, B.M.; Sutcliffe, M.J.; Munro, A.W.; Scrutton, N.S. Role of active site residues and solvent in proton transfer and the modulation of flavin reduction potential in bacterial morphinone reductase. *J. Biol. Chem.* **2005**, *280*, 27103–27110. [CrossRef] [PubMed]
115. Spiegelhauer, O.; Dickert, F.; Mende, S.; Niks, D.; Hille, R.; Ullmann, M.; Dobbek, H. Kinetic characterization of Xenobiotic reductase A from *Pseudomonas putida* 86. *Biochemistry* **2009**, *48*, 11412–11420. [CrossRef] [PubMed]
116. Pudney, C.R.; Hay, S.; Pang, J.Y.; Costello, C.; Leys, D.; Sutcliffe, M.J.; Scrutton, N.S. Mutagenesis of morphinone reductase induces multiple reactive configurations and identifies potential ambiguity in kinetic analysis of enzyme tunneling mechanisms. *J. Am. Chem. Soc.* **2007**, *129*, 13949–13956. [CrossRef] [PubMed]
117. Williams, R.E.; Rathbone, D.A.; Scrutton, N.S.; Bruce, N.C. Biotransformation of explosives by the old yellow enzyme family of flavoproteins. *Appl. Environ. Microbiol.* **2004**, *70*, 3566–3574. [CrossRef] [PubMed]
118. Reich, S.; Nestl, B.M.; Hauer, B. Loop-grafted old yellow enzymes in the bienzymatic cascade reduction of allylic alcohols. *ChemBioChem* **2016**, *17*, 561–565. [CrossRef] [PubMed]
119. Nestl, B.M.; Hauer, B. Engineering of flexible loops in enzymes. *ACS Catal.* **2014**, *4*, 3201–3211. [CrossRef]
120. Reich, S.; Kress, N.; Nestl, B.M.; Hauer, B. Variations in the stability of NCR ene reductase by rational enzyme loop modulation. *J. Struct. Biol.* **2014**, *185*, 228–233. [CrossRef] [PubMed]
121. Basran, J.; Harris, R.J.; Sutcliffe, M.J.; Scrutton, N.S. H-tunneling in the multiple H-transfers of the catalytic cycle of morphinone reductase and in the reductive half-reaction of the homologous pentaerythritol tetranitrate reductase. *J. Biol. Chem.* **2003**, *278*, 43973–43982. [CrossRef] [PubMed]
122. Pudney, C.R.; Hay, S.; Sutcliffe, M.J.; Scrutton, N.S. $\alpha$-Secondary isotope effects as probes of "tunneling-ready" configurations in enzymatic H-tunneling: Insight from environmentally coupled tunneling models. *J. Am. Chem. Soc.* **2006**, *128*, 14053–14058. [CrossRef] [PubMed]

123. Pudney, C.R.; Hay, S.; Levy, C.; Pang, J.Y.; Sutcliffe, M.J.; Leys, D.; Scrutton, N.S. Evidence to support the hypothesis that promoting vibrations enhance the rate of an enzyme catalyzed H-tunneling reaction. *J. Am. Chem. Soc.* **2009**, *131*, 17072–17073. [CrossRef] [PubMed]

124. Brenna, E.; Gatti, F.G.; Monti, D.; Parmeggiani, F.; Serra, S. Stereochemical outcome of the biocatalysed reduction of activated tetrasubstituted olefins by old yellow enzymes 1–3. *Adv. Synth. Catal.* **2012**, *354*, 105–112. [CrossRef]

125. Oberdorfer, G.; Steinkellner, G.; Stueckler, C.; Faber, K.; Gruber, K. Stereopreferences of old yellow enzymes: Structure correlations and sequence patterns in enoate reductases. *ChemCatChem* **2011**, *3*, 1562–1566. [CrossRef]

126. Classen, T.; Pietruszka, J.; Schuback, S.M. Revisiting the enantioselective sequence patterns in enoate reductases. *ChemCatChem* **2013**, *5*, 711–713. [CrossRef]

127. Pompeu, Y.A.; Sullivan, B.; Stewart, J.D. X-ray crystallography reveals how subtle changes control the orientation of substrate binding in an alkene reductase. *ACS Catal.* **2013**, *3*, 2376–2390. [CrossRef]

128. Walton, A.Z.; Sullivan, B.; Patterson-Orazem, A.C.; Stewart, J.D. Residues controlling facial selectivity in an alkene reductase and semirational alterations to create stereocomplementary variants. *ACS Catal.* **2014**, *4*, 2307–2318. [CrossRef] [PubMed]

129. Walton, A.Z.; Conerly, W.C.; Pompeu, Y.; Sullivan, B.; Stewart, J.D. Biocatalytic reductions of Baylis-Hillman adducts. *ACS Catal.* **2011**, *1*, 989–993. [CrossRef]

130. Amato, E.D.; Stewart, J.D. Applications of protein engineering to members of the old yellow enzyme family. *Biotechnol. Adv.* **2015**, *33*, 624–631. [CrossRef] [PubMed]

131. Bougioukou, D.J.; Kille, S.; Taglieber, A.; Reetz, M.T. Directed evolution of an enantioselective enoate-reductase: Testing the utility of iterative saturation mutagenesis. *Adv. Synth. Catal.* **2009**, *351*, 3287–3305. [CrossRef]

*catalysts*

MDPI

*Article*

# N-acetylglucosamine 2-Epimerase from *Pedobacter heparinus*: First Experimental Evidence of a Deprotonation/Reprotonation Mechanism

Su-Yan Wang [1], Pedro Laborda [1], Ai-Min Lu [2], Xu-Chu Duan [1], Hong-Yu Ma [3], Li Liu [1,*] and Josef Voglmeir [1,*]

[1]   Glycomics and Glycan Bioengineering Research Center (GGBRC), College of Food Science and Technology, Nanjing Agricultural University, Nanjing 210095, China; 2014208001@njau.edu.cn (S.-Y.W.); pedro.laborda@njau.edu.cn (P.L.); 2012208008@njau.edu.cn (X.-C.D.)
[2]   College of Sciences, Nanjing Agricultural University, Nanjing 210095, China; luaimin@njau.edu.cn
[3]   Department of Plant Pathology, Nanjing Agricultural University, Nanjing 210095, China; mahongyu@njau.edu.cn
*    Correspondence: lichen.liu@njau.edu.cn (L.L.); josef.voglmeir@njau.edu.cn (J.V.); Tel: +86-25-8439-9553 (L.L. & J.V.)

Academic Editors: Jose M. Palomo and Cesar Mateo
Received: 26 November 2016; Accepted: 12 December 2016; Published: 17 December 2016

**Abstract:** The control of cellular *N*-acetylmannosamine (ManNAc) levels has been postulated to be an effective way to modulate the decoration of cell surfaces with sialic acid. *N*-acetylglucosamine 2-epimerase catalyzes the interconversion of *N*-acetylglucosamine (GlcNAc) and ManNAc. Herein, we describe the cloning, expression, purification and biochemical characterization of an unstudied *N*-acetylglucosamine 2-epimerase from *Pedobacter heparinus* (PhGn2E). To further characterize the enzyme, several *N*-acylated glucosamine derivatives were chemically synthesized, and subsequently used to test the substrate specificity of PhGn2E. Furthermore, NMR studies of deuterium/hydrogen exchange at the anomeric hydroxy group and C-2 positions of the substrate in the reaction mixture confirmed for the first time the postulated epimerization reaction via ring-opening/enolate formation. Site-directed mutagenesis of key residues in the active site showed that Arg63 and Glu314 are directly involved in proton abstraction and re-incorporation onto the substrate. As all mechanistically relevant active site residues also occur in all mammalian isoforms, PhGn2E can serve as a model *N*-acetylglucosamine 2-epimerase for further elucidation of the active site mechanism in these enzymes.

**Keywords:** sialic acid metabolism; *N*-acetylglucosamine 2-epimerase; deprotonation/reprotonation mechanism; Neu5Ac analogues; synthesis of sialic acid analogues

## 1. Introduction

Sialic acids are naturally occurring carbohydrate derivatives of neuraminic acid and 2-keto-3-deoxy-D-glycero-D-galactonononic acid [1]. Sialic acids are generally found at the termini of protein and lipid linked glycoconjugates on cell surfaces, and have been demonstrated to play a crucial role in cell recognition, modulation of cell receptors, tumor metastasis and pathogen binding [2–6]. In recent years, remarkable efforts have been undertaken towards further understanding the biosynthesis and regulation of the sialic acid metabolism on a cellular level [7,8]. ManNAc has been postulated to be the main sialic acid precursor. The biosynthesis of *N*-acetylneuraminic acid (Neu5Ac) from ManNAc can be either achieved via phosphorylation by ManNAc 6-kinase and aldol addition of phosphoenolpyruvate (PEP) by sialic acid 9-P-synthase following an enzymatic dephosphorylation (Scheme 1, route A) [9,10], or the direct (reversible) aldol addition by sialic acid aldolase (Scheme 1, route B) [11]. The resulting sialic acid is then converted into its activated precursor form CMP-*N*-acetylneuraminic acid and transported into the Golgi apparatus, where it is ultimately utilized by various sialyltransferases to decorate oligosaccharide chains [12]. As ManNAc is the sole carbohydrate precursor in the biosynthesis of *N*-acetylneuraminic acid, a better understanding of the ManNAc biosynthesis will therefore also lead to a more profound knowledge of the sialic acid metabolism in cells.

The biosynthesis of ManNAc in mammals is regulated in two different ways (Scheme 1, top): in the liver, UDP-GlcNAc is epimerized to UDP-ManNAc and subsequently hydrolyzed to ManNAc by the bifunctional enzyme UDP-GlcNAc 2-epimerase [13]. Alternatively, GlcNAc can be directly epimerized to ManNAc by GlcNAc 2-epimerase (mainly in kidneys) [14,15]. Several GlcNAc 2-epimerases of either mammalian or bacterial origin have been functionally characterized so far [15–22]. The major difference between the two types of enzymes is that mammalian isoforms require nucleotides as co-factor, whereas bacterial isoforms may show enhanced activities in the presence of nucleotides but have no co-factor requirements [20,23].

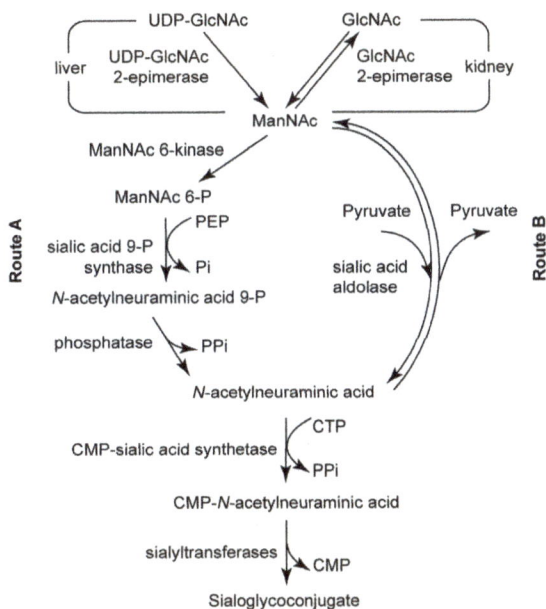

**Scheme 1.** Sialic acid biosynthesis.

GlcNAc 2-epimerases are interesting targets for the development of novel inhibitors of sialic acid biosynthesis. However, the main limitation in the development of effective inhibitors is the lack of information on the epimerization mechanism. Despite speculations by Samuel and Tanner that the mechanism of the closely related enzyme UDP-GlcNAc 2-epimerase is based on deprotonation/reprotonation or on elimination [24], no experimental studies have yet been reported to support either of these theories. Crystallographic studies of mammalian (porcine) and bacterial (*Anabaena* sp.) GlcNAc 2-epimerase apoproteins have indicated the involvement of various amino acid residues in the reaction mechanism [18,25]. Therefore, we performed a mutational analysis to further elucidate the role of these amino acids. Furthermore, we describe the purification and biochemical and mechanistic characterization of a novel GlcNAc 2-epimerase originating from the soil bacterium *Pedobacter heparinus*, confirming for the first time the postulated deprotonation/reprotonation mechanism for this class of enzymes.

## 2. Results

### 2.1. Cloning and Homology Analysis of PhGn2E Gene

A gene canditate encoding a putative GlcNAc 2-epimerase was selected from the genome of *Pedobacter heparinus*. The full length open reading frame (ORF) of the gene was successfully cloned and consisted of 1212 base pairs. A homology search revealed that PhGn2E is closely related to the GlcNAc 2-epimerase from mammals showing 35% identity to the human, bovine, murine and rat isoforms and 34% to the porcine isoform (Supplementary Figure S1). Among bacterial GlcNAc 2-epimerases, PhGn2E showed high amino acid sequence similarity with characterized isoforms from *Bacteroides ovatus* (48%) and *Anabaena* sp. (37%).

### 2.2. Protein Expression and Purification

The putative gene product was successfully expressed in soluble form. The recombinant protein containing an additional hexa-histidine tag at the C-terminus was purified to homogeneity as judged by SDS-PAGE (Figure 1A). The purified sample migrated as a single protein band with a molecular weight between 40 kDa and 50 kDa, which is in good agreement with the calculated mass of 47.1 kDa (including the 1.3 kDa hexa-histidine tag). The identity of the purified protein band was confirmed using tryptic peptide mass fingerprinting by matching the obtained MALDI-TOF MS data with the Mascot software database (Supplementary Table S1). The protein concentration of the purified PhGn2E was determined to be $2.73 \pm 0.05$ mg/mL and relative enzymatic activity was determined to be 3.59 U/mg (1 U is defined as the amount of enzyme which generates 1 µmol of *N*-acetylmannosamine in 1 min at 37 °C).

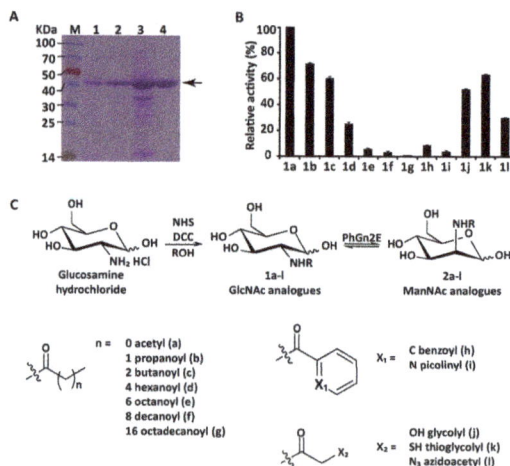

**Figure 1.** (**A**): SDS-PAGE analysis of heterologous expressed PhGn2E in *E. coli*. M: protein marker; 1: cell pellets before induction; 2: cell pellets after induction; 3: supernatant of cell lysate; 4: Ni-NTA purified enzyme; (**B**): Substrate promiscuity of PhGn2E. PhGn2E showed activity towards *N*-acetylglucosamine (**1a**); *N*-propanoylglucosamine (**1b**); *N*-butanoylglucosamine (**1c**); *N*-hexanoylglucosamine (**1d**); *N*-octanoylglucosamine (**1e**); *N*-decanoylglucosamine (**1f**); *N*-benzoylglucosamine (**1h**); *N*-picolinylglucosamine (**1i**); *N*-glycolylglucosamine (**1j**); *N*-thioglycolylglucosamine (**1k**) and *N*-azidoacetylglucosamine (**1l**). PhGn2E showed no activity towards *N*-octadecanoylglucosamine (**1g**). Data are presented as mean values $\pm$ standard deviation of three independent experiments; (**C**): Reaction scheme of the synthesis of GlcNAc analogs by treatment with NHS (*N*-hydroxysuccinimide), DCC (Dicyclohexylcarbodiimide) and the corresponding acid using glucosamine as the starting material and epimerization to ManNAc analogs using PhGn2E yielding ManNAc analogues (**2a–2l**).

## 2.3. Substrate Specificity

The screening of various chemically synthesized substrates (Figure 1B, Supplementary Figures S2 and S3) revealed that recombinant PhGn2E was able to catalyze the epimerization reaction when GlcNAc derivatives bearing small *N*-acyl groups such as *N*-propanoylglucosamine, *N*-butanoylglucosamine, *N*-hexanoylglucosamine, *N*-octanoylglucosamine, *N*-glycolylglucosamine or *N*-thioglycolyl- glucosamine (Figure 1C, compounds **1b**, **1c**, **1d**, **1e**, **1j** and **1k**, respectively). However, the activity decreased significantly when substrates with bulkier modifications such as *N*-decanoylglucosamine and *N*-octadecanoylglucosamine were used (Figure 1C, compounds **1f** and **1g**, respectively). PhGn2E was also able to catalyze the epimerization reaction of GlcNAc derivatives bearing a hydroxyl (**1j**), a thiol (**1k**) or azido (*N*-azidoacetylglucosamine, **1l**) moiety in the *N*-linked substituent. Low, but detectable activities towards GlcNAc derivatives containing aromatic side-chains (*N*-benzonylglucosamine, **1h** and *N*-picolinylglucosamine, **1i**) were also observed. Both NMR analysis and platereader based activity assays indicated that PhGn2E showed no activity towards glucosamine, α-methyl GlcNAc, β-methyl GlcNAc and UDP-GlcNAc. Furthermore, all the synthesized GlcNAc derivatives except **1g** could be successfully converted into sialic acid analogues and linked onto X-gal following the synthetic route shown in Figure 2 and Supplementary Figures S4–S6.

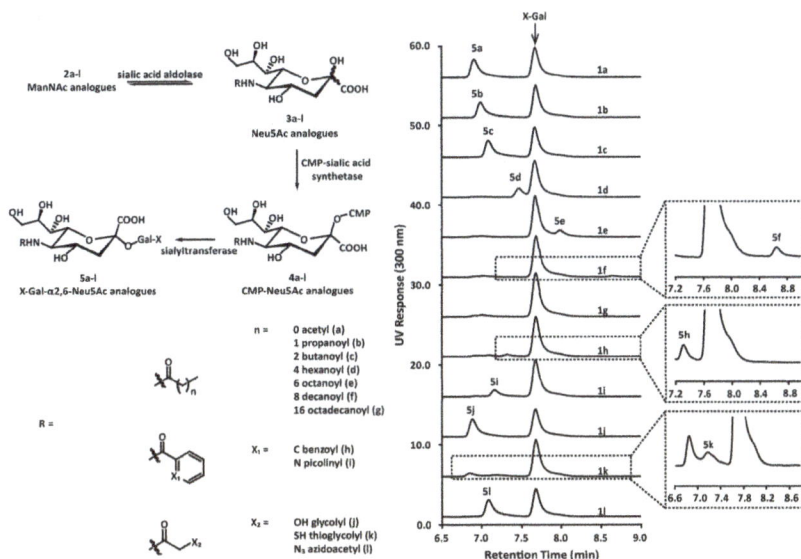

**Figure 2.** Reaction scheme of the synthesis route of X-gal sialosides. **2a–l** was achieved by PhGn2E using **1a–l** as substrates. The resulting X-gal sialosides (**5a**, **5b**, **5c**, **5d**, **5e**, **5f**, **5h,5i**, **5j**, **5k** and **5l**) were synthesized by a four-enzyme reaction starting with *N*-acetylglucosamine (**1a**); *N*-propanoylglucosamine (**1b**); *N*-butanoyl- glucosamine (**1c**); *N*-hexanoylglucosamine (**1d**); *N*-octanoylglucosamine (**1e**); *N*-decanoylglucosamine (**1f**); *N*-benzonylglucosamine (**1h**); *N*-picolinylglucosamine (**1i**); *N*-glycolylglucosamine (**1j**); *N*-thioglycolylgluco- samine (**1k**) and *N*-azidoacetylglucosamine (**1l**) as initial substrates with relative conversion rates of 37%, 33%, 37%, 15%, 16%, 2.3%, 3.1%, 13%, 39%, 3.2%, and 38%, respectively. No yield from **1g** was observed to obtain **5g**.

## 2.4. Biochemical Characterization

PhGn2E activity was tested in a coupled enzymatic assay, by monitoring the *N*-acyl-D-mannosamine dehydrogenase-catalyzed reduction of NAD⁺ in the presence of ManNAc to NADH, which can be followed in a plate reader-based photometric assay. The optimum reaction temperature was determined to be 37 °C (Figure 3A). Above 37 °C, enzymatic activity rapidly decreased with increasing temperatures. The enzyme showed reasonable thermal stability when incubated at 45 °C or below for 2 h, and showed no loss of activity when incubated at 30 °C or below for 24 h (Supplementary Figure S7). The enzyme showed good activity in a relatively broad pH range from pH 7.0 to pH 10.0, but decreased rapidly below and above this range (Figure 3B). Considering that small proportions of GlcNAc may also epimerize into ManNAc under stronger basic conditions, pH 8.0 was chosen for further experiments. The addition of $Ca^{2+}$, $Mg^{2+}$, $Mn^{2+}$, $Zn^{2+}$, $Cu^{2+}$, $Co^{2+}$ or EDTA had no effect on the enzyme's activity, while $Fe^{2+}$, $Fe^{3+}$ and $Al^{3+}$ slightly inhibited and $Ni^{2+}$ slightly increased the activity of PhGn2E (Figure 3C). Denaturants (urea and 2-mercarptoethanol) and detergents (SDS and Triton X-100) showed significant inhibitory effects on the enzymatic activity (Supplementary Figure S8). The effect of nucleotides on PhGn2E was also examined (Supplementary Figure S9A), showing that ATP but no other nucleotide enhanced the activity of PhGn2E. To evaluate the influence of adenine-derived nucleotides on PhGn2E activity, the inhibition of AMP, ADP and ATP was further studied in a dose-dependent manner (Supplementary Figure S9B). Whereas AMP and ADP had no significant effect at any measured concentration, the addition of 100 μM of ATP resulted almost in a doubling of enzymatic activity and the addition of 1 mM of ATP led to a 4-fold increase. The $K_M$ value of PhGn2E was 82 ± 7 mM for GlcNAc, the $V_{max}$ value 197 ± 4 μM·min$^{-1}$ and the $k_{cat}$ value 340 ± 3 min$^{-1}$. A mutational analysis of PhGn2E was performed in order to confirm the residues which are involved in the epimerization reaction. Arg63, His244, Glu314 and His378 were chosen as the target amino acids for this study. The activities of PhGn2E mutants H244A and H378A were 13.6% ± 1.0% and 6.9% ± 0.9% of that of the wild type enzyme, whereas the mutants R63A and E314A showed no epimerase activity (Figure 3D).

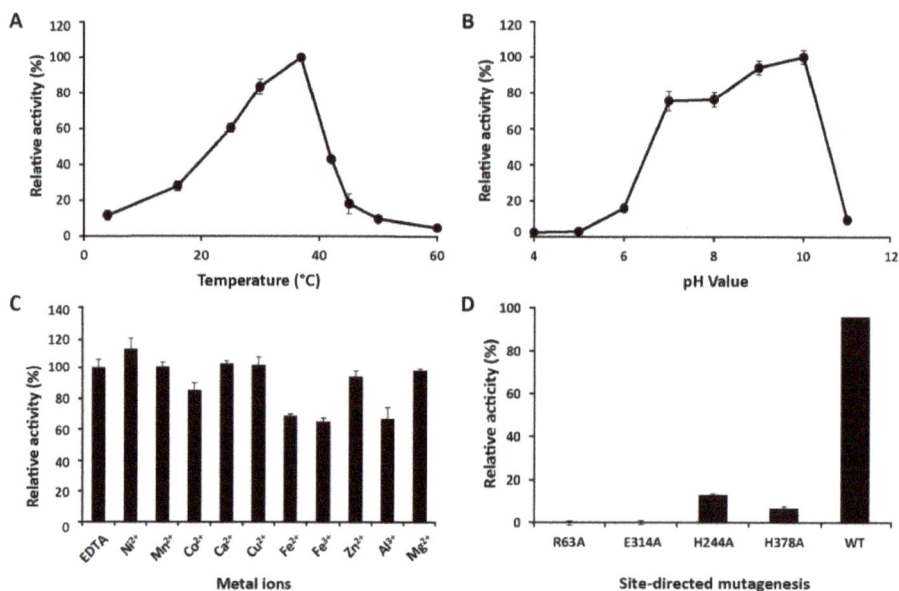

**Figure 3.** Biochemical characterization of PhGn2E. (**A**): Temperature dependency of recombinant PhGn2E; (**B**): pH-dependency of recombinant PhGn2E; (**C**): Impact of metal ions on the enzymatic activity of PhGn2E; (**D**): Activity study of PhGn2E active site mutants. Negative controls contained heat-inactivated wild-type (WT) PhGn2E. Data are presented as mean values ± standard deviation of three independent experiments.

## 2.5. Hydrogen/Deuterium Exchange Analysis

[1]H-, [13]C-, TOCSY-, and HSQC-NMR spectra of GlcNAc and ManNAc provided by the BMRB repository [26] were used for the accurate assignment of the obtained NMR signals. After the epimerization reaction in deuterated water (D$_2$O), the acquired [1]H-NMR spectrum revealed the disappearance of the signal corresponding to the C-2 linked proton of ManNAc (ManNAc H2, 4.5–4.3 ppm) indicating a hydrogen/deuterium exchange at this position (Figure 4A–D). Similarly, the relative integral value of the multiplet between 4.1 and 3.3 ppm (protons linked to C-2, C-3, C-4, C-5 and C-6) of the reaction mixture showed a ratio of 5:1 with respect to the anomeric proton signals (4 signals: 5.19, 5.12, 5.02 and 4.71 ppm) instead of the expected ration of 6:1, which is in agreement with hydrogen/deuterium exchange at C-2 of GlcNAc (GlcNAc H2, 4.1–3.3 ppm) (Figure 4D). On the other hand, a deuterium/hydrogen exchange could be also observed in the anomeric alcohol of GlcNAc after the addition of heat-inactivated PhGn2E. However, integrating the relative areas of the signals corresponding to the anomeric proton of the β anomer (βH1, 6.44 ppm) and the anomeric alcohol of the β anomer (βOH1, 4.93 ppm) showed 6% of deuterium/hydrogen exchange when GlcNAc was incubated for 15 min using D$_2$O as the solvent in the presence of the inactivated epimerase (Figure 5A–C); while the almost complete disappearance of the signal corresponding to the anomeric alcohol was observed when active PhGn2E was added (Figure 5D).

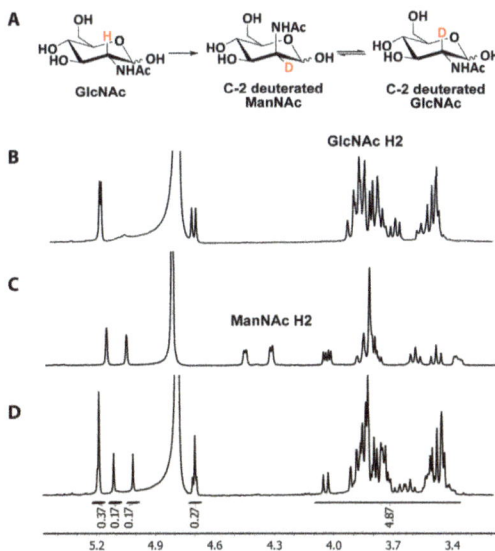

**Figure 4.** (**A**): Reaction scheme of PhGn2E-catalyzed deuterium substitution at the C-2 position of GlcNAc allowing the formation C-2 deuterated ManNAc; (**B**): [1]H-NMR spectrum of GlcNAc in D$_2$O; (**C**): [1]H-NMR spectrum of ManNAc in D$_2$O; (**D**): [1]H-NMR spectrum of GlcNAc after incubation with PhGn2E in D$_2$O.

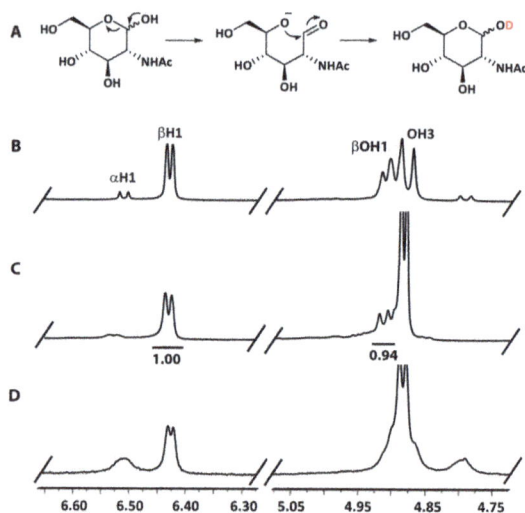

**Figure 5.** (**A**): Reaction scheme of the hydrogen/deuterium exchange on the anomeric alcohol of GlcNAc. The pyranose ring opening involves the formation of a double bond at the anomeric carbon to form an intermediate aldehyde. The C-5 oxygen of the intermediate then attacks the anomeric carbon, closing the ring again. The anomeric oxygen is deuterated by the solvent; (**B**): $^1$H-NMR spectrum of GlcNAc in DMSO-$d_6$; (**C**): $^1$H-NMR spectrum of the reaction mixture containing 20 mM GlcNAc and inactivated PhGn2E in D$_2$O after 15 min incubation at 37 °C (spectrum was recorded in DMSO-$d_6$); (**D**): $^1$H-NMR spectrum of the reaction mixture containing 20 mM GlcNAc and PhGn2E in D$_2$O after 15 min incubation at 37 °C (spectrum was recorded in DMSO-$d_6$).

## 3. Discussion

### 3.1. Characterization

The recombinant form of PhGn2E could be successfully purified and was observed as a single protein band of expected molecular weight on SDS-PAGE. The enzyme showed good expression levels and had, after purification, a high specific activity of 3.59 U/mg. Furthermore, microplate reader-, HPLC-, and NMR-based methods unambiguously demonstrated the enzymatic epimerization activity of PhGn2E.

Although PhGn2E showed no activity at temperatures of 50 °C and above, the enzyme showed reasonable stability when incubated at 37 °C for up to 12 h. This makes PhGn2E potentially suitable for biotechnological applications at ambient temperatures with longer incubation times, such as continuous biotransformations. The optimum temperature of PhGn2E was determined to be 37 °C. This value is comparable with the reported temperature optimum for the cyanobacterium *Synechocystis* sp. PCC 6803 [27], whereas *B. ovatus*, porcine, and *A. variabilis* isoforms showed higher temperature optima (45 to 60 °C, respectively) [19,28,29]. The pH optimum of PhGn2E lies slightly higher than other characterized epimerases, which were reported to lie between 6.8–8.5 for the porcine, *Anabaena* sp., *B. ovatus*, and *Synechocystis* sp. PCC6803 homologues [22,23,27,29]. The addition of EDTA or metal ions did not significantly decrease or inhibit PhGn2E activity, which indicates that metal ions are not involved in the catalytic mechanism of the enzyme. Kinetic analysis showed that the measured $K_M$ value of 82 mM was higher compared to other reported $K_M$ values (between 7 and 32 mM) [22,23,27,29,30], which might be an advantage for biotransformations performed at high substrate concentrations.

While several bacterial GlcNAc 2-epimerases show activity in absence of nucleotides, the addition of ATP has been reported to enhance enzymatic activity for some isoforms. For example, the activity of *Anabaena* sp. epimerase was enhanced in the presence of ATP and the non-hydrolyzable γ-phosphate analogue 5′-adenylyl imidodiphosphate [23]. It has been demonstrated that the addition of ATP stabilizes the *B. ovatus* epimerase leading to an increased denaturation temperature and an enhanced pH activity range [31]. In a similar manner, the activity of PhGn2E also increased significantly in the presence of ATP. While mammalian epimerases can be activated in presence of a wide range of nucleotides such as AMP, ADP, ATP, CTP, GTP or UTP [30], PhGn2E activity remained unchanged in the presence of these nucleotides. Two different theories of how ATP-binding effects the activity of GlcNAc 2-epimerase currently exist; Liao et al. suggested an ATP-binding site in *Anabaena* sp. GlcNAc 2-epimerase site 151-KDNPKGKYTK-160 (Supplementary Figure S1), and that the residues Lys151 and Lys160 are required for ATP binding [23]. A second hypothesis by Sola-Carvajal et al. suggests that the lysine residues in a different motif located at the *C*-terminus of the analyzed *B. ovatus* isoform (392-KGGKWKG-398, Supplementary Figure S1) are responsible for ATP binding [31]. This motif was also found in the *C*-terminal region of PhGn2E (369-KGNLFKG-375), and in other functionally described isoforms from *Anabaena* sp. and mammals (Supplementary Figure S1), suggesting that the lysine residues at the *C*-terminal region may be responsible for the binding of ATP.

*3.2. Chemoenzymatic Synthesis of Mannosamine and Sialic Acid Derivatives*

The synthesis of mannosamine and sialic acid derivatives is one of the main limitations in studying them as potential inhibitors of the sialic acid metabolic pathway [32–37]. In this field, several research groups have reported the synthesis of mannosamine derivatized with *N*-linked aliphatic chains. This is generally achieved by using NHS-activated acids [38,39], or alternatively, can also be achieved by treatment of isobutyl chloroformate with an acid, yielding an amine-reactive anhydride [36,40]. Sampathkumar et al. reported the synthesis of peracetylated ManNAc analogues by treatment with naphthaldehyde for *N*-activation using ManN hydrochloride as starting material [41]. However, the high price of mannosamine is a major limitation in the synthesis of sialic acid analogues. Incorporating PhGn2E into the synthetic route allows the use of glucosamine-based analogues, which reduces the cost of the synthesis to a fraction of that of the mannosamine-based synthesis.

The mannosamine derivatives obtained from the PhGn2E-catalyzed reaction were used as precursors for the synthesis of Neu5Ac (**3a**) analogues. These Neu5Ac analogues bearing alternative *N*-linked acyl chains could be synthetized by the sialic acid aldolase-catalyzed addition of pyruvate. Several chemical strategies to generate Neu5Ac derivatives with C-5 modifications have been successfully applied so far, including modification with benzyl- and azido groups [42,43]. A free amino group has been demonstrated to be a suitable nucleophile for protection with Fmoc, Alloc or Boc carbamates [42,43]. Unprotected neuraminic acid had also been used for the addition of trifluoroacetamido, trichloroethoxycarbonyl or 4-*O*-oxazolidinone groups, replacing the acetyl group [43]. Keppler et al. used a whole cell biotransformation to produce *N*-propanoyl, *N*-butanoyl and *N*-pentanoyl neuraminic acids by fermentation in Madin-Darby canine kidney cells using the corresponding ManNAc derivatives as the starting material [44]. The same strategy was applied by Lundgren et al. to produce a sialic acid derivative from *N*-butanoylglucosamine in genetically engineered *E. coli* [45]. The chemoenzymatic synthesis of various sialosides based on mannosamines derivatized with fluoride-, azido- or methoxide groups in a one-pot, three-enzyme approach with recombinant enzymes was later reported by the group of Chen [38,46]. The herein characterized PhGn2E epimerase may also be easily integrated into reported synthetic procedures, which would allow the use of glucosamine derivatives as a starting material.

*3.3. Mechanistic Studies*

Recorded NMR spectra showed hydrogen/deuterium exchange at C-2 in both GlcNAc and ManNAc when the reaction was performed in $D_2O$. This result can be explained considering that the enzyme is catalyzing the deprotonation of GlcNAc at C-2 and, at the same time, the incorporation of deuterium on the opposite face, promoting the epimerization of GlcNAc to C-2 deuterated ManNAc (Figure 4A). Four different mechanisms have been described for epimerization reactions in sugars: the mutarotation mechanism, the transient oxidation-reduction mechanism, the elimination mechanism and the deprotonation/reprotonation mechanism (Supplementary Figure S10) [24]. The mutarotation mechanism is only applicable for epimerization reactions at the anomeric center and can be therefore excluded. As GlcNAc 2-epimerases contain no co-enzymes such as FAD or NAD+, a transient oxidation-reduction mechanism, with no observable proton/deuterium exchange in the substrate, can be also excluded as possible mechanism. In both the deprotonation/reprotonation mechanism and the elimination mechanism, two different residues are involved in the proton capture and proton donation steps, allowing hydrogen/deuterium exchange. To discern between the two possible mechanisms, the activity of PhGn2E towards anomeric O-substituted GlcNAc derivatives was studied. PhGn2E showed no activity towards methyl α-GlcNAc, methyl β-GlcNAc or UDP-GlcNAc, indicating that the enzyme is not capable of catalyzing the epimerization reaction in closed ring formation. In addition, deuterium exchange on the anomeric alcohol is faster in presence of PhGn2E (Figure 5C,D). According to the described mechanism for UDP-GlcNAc 2-epimerase, an elimination mechanism for the GlcNAc/ManNAc enzymatic epimerization would involve the release and subsequent incorporation of the hydroxyl group [47], which would consequently lead to no hydrogen/deuterium exchange in the anomeric alcohol. Thus, the hydrogen/deuterium exchange is produced via pyranose ring opening and closure (Figure 5A). This result is in agreement with the deprotonation/reprotonation mechanism wherein the ring opening occurs in order to stabilize by enolate intermediate the anion formed in C-2 after the abstraction of the proton (Figure 6), confirming that the catalytic activity of PhGn2E is based on a deprotonation/reprotonation mechanism.

The deprotonation/reprotonation mechanism requires the participation of two proximate amino acid residues in the active site of the epimerase [24]. Itoh et al. obtained the crystal structure of the porcine isoform, and showed that four amino acid residues (Arg60, His248, Glu251 and His382) are close enough to the active center to be potentially important for catalytic function [18]. While the X-ray diffraction of the enzyme was performed in presence of GlcNAc, the resolution of the substrate was not high enough to carry out further crystallographic refinements and, for this reason, the exact residues involved in the proton abstraction/donation steps could not be identified. Lee et al. reported the three-dimensional structure of GlcNAc 2-epimerase from *Anabaena* sp., and that the substitution of amino acids in the active site using site-directed mutagenesis showed that Arg57, His239, Glu308, His372, are essential for the epimerization reaction [25]. The sequence alignment of PhGn2E, porcine and *Anabaena* sp. isoforms (Supplementary Figure S1) show high similarities, especially close to the above mentioned amino acids surrounding the active site. In order to identify the two residues involved in the proton capture/donation steps, the activity of PhGn2E mutants was further investigated. Despite the results of Lee et al., which suggest that His244 and H378 should be responsible for the deprotonation/reprotonation mechanism [25], the PhGn2E mutants H244A and H378A could efficiently catalyze the epimerization reaction from GlcNAc to ManNAc. On the other hand, the PhGn2E mutants E314A and R63A were not capable of catalyzing the epimerization reaction indicating that Arg63 and His372 are the residues involved in the proton abstraction/donation steps (Figure 3D). Thus, one of these residues must participate in abstracting the hydrogen linked to C-2 of GlcNAc followed by incorporation of the hydrogen in the same carbon but in the opposite face by the other residue producing the inversion of configuration.

In order to assign the role of each residue, hydrogen/deuterium exchange studies of PhGn2E mutants E314A and R63A were performed. However, no hydrogen/deuterium exchange could be observed using either ManNAc or GlcNAc as initial substrate. Reports on the *B. ovatus* epimerase indicated that the epimerization reaction from GlcNAc to ManNAc is faster than the epimerization reaction from ManNAc to GlcNAc [22]. The reason for this may be pKa values of arginine (pKa ≈ 12.5) and glutamic acid (pKa ≈ 3.1) residues, which indicate that a high proportion of protonated arginine and deprotonated glutamic acid residues exist at physiological pH conditions. Based on these preconditions, we can propose that a GlcNAc in open conformation enters in the central cavity between Arg63 and Glu314. Then, Glu314 deprotonates C-2 of GlcNAc while Arg63 incorporates the proton on the same carbon but on the opposite face yielding ManNAc (Figure 6). Due to the reversibility of the mechanism, GlcNAc 2-epimerase can also catalyze the epimerization from ManNAc to GlcNAc through the deprotonation of ManNAc in C-2 by Arg63 followed by the reprotonation by Glu314 to obtain GlcNAc. This makes GlcNAc 2-epimerases unique among deprotonation/reprotonation epimerases which commonly employ residues containing carbonyl, carboxylic acid and ester groups as key amino acids for the reaction [24].

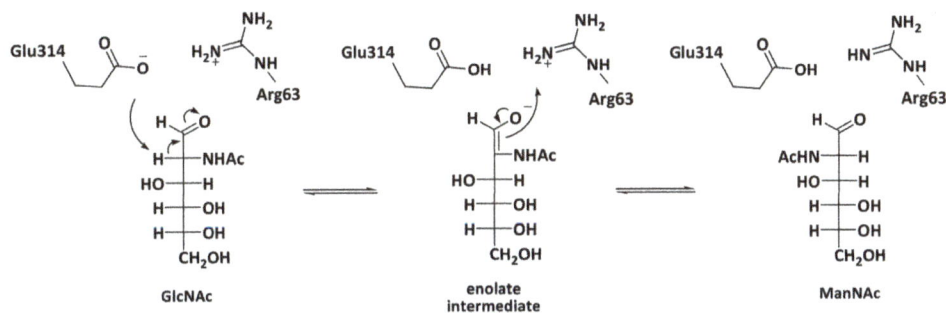

**Figure 6.** Deprotonation/reprotonation mechanism model of PhGn2E. Glu314 is able to deprotonate and protonate C-2 of GlcNAc while Arg63 is the responsible of the deprotonation and protonation in C-2 of ManNAc.

PhGn2E is highly homologous with mammalian GlcNAc 2-epimerases. Arg63 and Glu314 can be found in the human, bovine, murine, porcine and rat GlcNAc 2-epimerase sequences (Supplementary Figure S1). Furthermore, the crystal structure of porcine GlcNAc 2-epimerase showed that the corresponding arginine and glutamic acid residues are also symmetrically located in the active site (Supplementary Figure S11) [18]. The existence of these homologous residues in the active site of the porcine epimerases, as well as a high similarity in the surrounding residues, is a strong indication that mammalian epimerases also employ a deprotonation/reprotonation mechanism to carry out the C-2 inversion.

## 4. Materials and Methods

### 4.1. General

DNA polymerases were purchased from Takara (Dalian, China); Restriction endonucleases and T4 ligases were obtained from Thermo Fisher Scientific (Shanghai, China); DNA Gel Purification and Plasmid Extraction kits were from Axygen (Beijing, China). Oligonucleotide primers were purchased from GenScript (Nanjing, China). Buffers, salt and other chemicals used in this study were purchased at the highest grade from various suppliers.

### 4.2. Bacterial Strains

*Pedobacter heparinus* (DSM 2366) was obtained from the German Collection of Microorganisms and Cell Cultures (DSMZ, Darmstadt, Germany). *E. coli* Mach 1 cells (Life Technologies, Shanghai, China) were used for plasmid amplification and manipulation. *E. coli* BL21 (DE3) cells (Invitrogen, Shanghai, China) were used for protein expression experiments.

### 4.3. Gene Cloning and Construction of the Expression Vector

Genomic DNA was isolated from stored lyophilized culture samples of *Pedobacter heparinus* according to the method described by Mahuku [48]. The oligonucleotide primers for amplifying a putative GlcNAc 2-epimerase were designed based on the *P. heparinus* genome information provided by the Pathosystems Resource Integration Center (PATRIC) [49] (ph.ggbrc.com) as follows: 5′-GGAATTCCATATGGTTATAGAATATACATTAGAAAAAT-3′ (forward primer) and 5′-CCGCTCGAGTTGAGCCGTATAGGAAGCA-3′ (reverse primer). Gene amplification was carried out using Primestar HS DNA polymerase based on the manufacturer's instructions. Briefly, the PCR amplification was performed using 35 PCR cycles consisting of denaturation at 95 °C for 10 s annealing at 55 °C for 30 s, and elongation at 72 °C for 1 min. The PCR fragments were purified on an agarose gel, digested with the restriction endonucleases Nde I and Xho I and subsequently ligated into a pET-30a expression vector, which was pre-digested using the same restriction endonucleases. The ligation mixtures were transformed into *E. coli* Mach1 T1 competent cells and were selected on Luria-Bertani (LB) agar containing 50 µg/mL kanamycin. Transformants containing the expected plasmid construct were screened by Sanger DNA sequencing (Genscript, Nanjing, China). A clone containing the expected plasmid construct was stored at −80 °C and used for further experiments. The extraction of plasmids, endonuclease treatments, ligation, DNA purification and transformation procedures were carried out using standard protocols.

### 4.4. Expression and Protein Purification

*E. coli* BL 21 (DE3) cells were transformed with the expression plasmid bearing the PhGn2E gene. A single colony was transferred into 5 mL of LB medium containing 50 µg/mL kanamycin for overnight shaking at 37 °C. The cells were then transferred into 400 mL LB medium and shaken (200 rpm) at 37 °C until the optical density at 600 nm ($OD_{600}$) reached a value between 0.5 and 0.8. Recombinant protein expression was induced by adding 1 mM IPTG to the fermentation broth. After further 24 h of shaking at 18 °C, cells were harvested by centrifugation (15 min at 5000 $g$), re-suspended in lysis buffer (50 mM Tris, 100 mM NaCl, 1% ($w/v$) Triton X-100, adjusted to pH 8.0 with HCl) and disrupted by sonication (40 on/off cycles with 20 µm probe amplitude for 15 s). The supernatant of the cell lysate was collected (20 min centrifugation at 14,000 $g$), and loaded onto a $Ni^{2+}$-nitrilotriacetate agarose affinity column (2 mL bed volume, Qiagen, Shanghai, China) equilibrated with washing buffer (50 mM Tris/HCl, 50 mM NaCl, pH 8.0). After washing the column with 30 mL of washing buffer, the bound target protein was then eluted in 1 mL fractions of elution buffer (50 mM Tris/HCl, 50 mM NaCl, 500 mM imidazole, pH 8.0). The progress of protein expression, cell lysis and protein purification was analyzed by SDS-polyacrylamide gel electrophoresis (SDS-PAGE) and visualized using Coomassie brilliant blue G-250. Elution fractions showing the highest purity were pooled and stored in 30% glycerol ($w/v$) at −80 °C for further experiments. The protein concentration of this pool was measured using a Bradford protein quantification kit (Sangon Biotech, Shanghai, China). Furthermore, Coomassie-stained gels of purified proteins were subject to in-gel tryptic digest and subsequent MALDI-TOF mass spectrometric analysis (Bruker Ultraflex Extreme, Bremen, Germany).

### 4.5. Enzymatic Assay

PhGn2E activity was detected based on the method described by Sola-Carvajal et al. [22] with slight modifications. In general, 10 µL assays containing 400 mM of GlcNAc and 0.58 µM of PhGn2E in citrate/phosphate buffer (50 mM, pH 8.0) and were incubated at 37 °C for 5 min. The epimerase activity was quenched by adding 70 µL of cold methanol and the protein was denatured by storing the samples for 2 h at −80 °C. After solvent evaporation, 80 µL of deionized $H_2O$ were added to the samples. The formation of ManNAc was analyzed using 3.4 µM of commercial *N*-acyl-D-mannosamine dehydrogenase (Qlyco, Nanjing, China) and $NAD^+$ (2 mM) as a co-factor in a continuous platereader assay at 37 °C (Thermo Multiscan FC, Shanghai, China). An increase in absorbance at 340 nm resulted from the reduction of $NAD^+$ to NADH when ManNAc was enzymatically oxidized to the corresponding lactone.

*4.6. Biochemical Characterization*

The temperature optimum of PhGn2E was determined by incubating reaction mixtures at various temperatures between 4 °C to 70 °C and the generation of ManNAc was subsequently determined with the assay conditions described above. The pH optimum was investigated using a series of citric acid/phosphate buffers (50 mM) ranging from pH 4.0 to pH 10.5. The thermal stability of PhGn2E was examined by incubating the enzyme at temperatures between 22 °C and 52 °C for various durations. Negative controls were prepared without the presence of enzyme. To evaluate the effect of metal ions on the activity of PhGn2E, epimerization reactions were performed in the presence of either 1 mM $Ca^{2+}$, $Mg^{2+}$, $Mn^{2+}$, $Zn^{2+}$, $Cu^{2+}$, $Co^{2+}$, $Fe^{2+}$, $Fe^{3+}$, $Ni^{2+}$, $Al^{3+}$ (all in their chloride form) or 1 mM ethylenediaminetetraacetate (EDTA).

The effect of several denaturants and detergents on PhGn2E was assessed for 2-mercaptoethanol (10 mM, 50 mM, 100 mM), urea (0.5 M, 1 M, 2 M), Triton-X 100 (0.1%, 0.5%, 1%) and SDS (0.1%, 0.5%, 1%). A series of nucleotides (AMP, ADP, ATP, CMP, CTP, GMP, GDP, GTP, UTP) was added in various concentrations to investigate their effect on the recombinant PhGn2E. Kinetic parameters were determined by using different concentrations of GlcNAc (ranging between 50 mM and 1200 mM). $V_{max}$, $K_M$ and $k_{cat}$ values were calculated by applying a non-linear regression model using Labplot data analysis software (Version 2.0.1.) (labplot.kde.org).

*4.7. Synthesis of N-Substituted Glucosamine Derivatives and Determination of PhGn2E Substrate Promiscuity*

GlcNAc analogues bearing different *N*-linked substituents were synthesized using a similar activated ester method (AES) as described previously [50]. In brief, 1 mmol of *N*-hydroxysuccinimide (NHS), 1 mmol dicyclohexylcarbodiimide (DCC) and 1 mmol of acid (acetic acid, propionic acid, butyric acid, hexanoic acid, octanoic acid, decanoic acid, stearic acid, benzoic acid, 2-picolinic acid, glycolic acid, thioglycolic acid, or azidoacetic acid) were dissolved in ethyl acetate (6 mL) and stirred overnight at room temperature. The reaction mixture was centrifuged at 14,000 *g* for 10 min. The supernatant was added drop-wise into 4 mL of a glucosamine solution (0.8 mmol GlcN in 10% (*v*/*v*) methanolic triethylamine). The resulting solution was stirred for 2 h at room temperature, the solvent was evaporated under reduced pressure and the resulting solid re-dissolved in 1.6 mL of an aqueous methanol solution (10% *v*/*v*). All substrates showed good solubility, with the exception of *N*-octadecanoylglucosamine (solubility of 8.8 mg/mL in 10% aqueous methanol at 37 °C). The substrate promiscuity of PhGn2E was measured using the platereader-based assay described in Section 4.5.

NMR analysis was used to evaluate enzyme-catalyzed reactions towards glucosamine, α-methyl GlcNAc and β-methyl GlcNAc. $^1$H-NMR spectra were acquired on a 400 MHz NMR instrument (Bruker, Bremen, Germany) using deuterium oxide as the solvent.

*4.8. Enzymatic Synthesis and Analysis of Indoxylsialosides*

Sialic acid analogues were synthesized based on a method previously reported by Cao et al. for *p*-nitrophenyl-linked sialosides [38]. This one-pot, four-enzyme reaction (Scheme 1, Route B) consists of the C2-epimerization of the GlcNAc derivatives, followed by an aldolase catalyzed aldol reaction to obtain C5-derivatized neuraminic acids. These derivatives were then activated using CMP-sialic acid synthetase and transferred to the acceptor substrate 5-bromo-4-chloro-3-indolyl-β-D-galactose (X-gal) by a sialic acid transferase. The expression and purification of the three latter enzymes was performed as previously described [51]. The reaction mixture consisted of 50 mM GlcNAc analogue, 2 mM CTP, 2 mM $MgCl_2$, 10 mM pyruvate, 2 mM X-gal, 50 mM MES (pH 6.5), 3 μM PhGn2E, 3 μM *E. coli* aldolase, 3 μM *Neisseria meningitidis* CMP-sialic acid synthetase and 5 μM *Photobacterium damsalae* sialic acid transferase. After 16 h of incubation at 37 °C, samples were analyzed using reversed phase HPLC with online UV and ESI-MS detection (Shimadzu MS-2020, Tokyo, Japan). The analytes were separated using a HyperClone 5 μm ODS column (250 × 4.6 mm, Phenomenex), with an ammonium formate eluent (50 mM, pH 4.5, 1 mL/min) mixed with increasing concentrations of acetonitrile (from 10% to 60% in the first 8 min) and detected at a UV wavelength of 300 nm. The ESI-mass detection was performed in negative ionization mode scanning an m/z range between 300 and 1000.

*4.9. Mechanistic NMR Studies*

$^1$H-NMR spectra were recorded on a Bruker AV-400 instrument, using the residual deuterium oxide and dimethyl sulfoxide-$d_6$ (DMSO-$d_6$) residual solvent signals as internal standards. For studying the C-2 hydrogen/deuterium exchange, approx. 3 mg of purified PhGn2E were dialyzed for 72 h against deionized water and subsequently lyophilized. The dried enzyme was re-dissolved in 600 μL of deuterium oxide containing 100 mM GlcNAc and incubated at 37 °C overnight. Spectra were recorded without any further sample treatment. For the hydrogen/deuterium exchange analysis of the anomeric alcohol, 3 mg of lyophilized epimerase was dissolved 300 μL of deuterium oxide containing 20 mM GlcNAc, and incubated at 37 °C for 15 min. After the solvent was evaporated, $^1$H-NMR spectra were collected in deuterated DMSO-$d_6$. Negative control experiments were performed by inactivating the lyophilized epimerase with methanol.

*4.10. Activity Screening Using PhGn2E Mutants*

Complementary oligonucleotide primers containing the desired mutation were used to generate PhGn2E mutant genes (Supplementary Table S2) according to the QuikChange overlap PCR Site-Directed Mutagenesis protocol (Stratagene). Verified plasmids containing the desired mutation in the target gene were transformed into *E. coli* BL21 (DE3) competent cells. PhGn2E mutant proteins were expressed and purified as described in Section 4.4. Activity tests were performed in 40 μL reaction mixtures containing 50 mM GlcNAc, 3 μM PhGn2E mutant or wild type in citrate/phosphate buffer (50 mM, pH 8.0), and were incubated at 37 °C for 4 h. Negative controls contained heat-inactivated PhGn2E (95 °C for 15 min) instead of the native enzyme. The reaction mixture was then analyzed using the platereader-based assay described in Section 4.5.

## 5. Conclusions

Our study describes the biochemical characterization and the substrate promiscuity of a previously unstudied GlcNAc 2-epimerase from *Pedobacter heparinus*. For the first time, experimental evidence of the proposed deprotonation/reprotonation mechanism for this type of enzyme could be demonstrated. This was achieved in NMR experiments by observing the hydrogen/deuterium exchange at C-2 and in the anomeric alcohol of the substrates. An activity study of PhGn2E mutant enzymes suggests that the proton abstraction/addition is catalyzed by Arg63 and Glu314. These residues do also appear in the active site of the closely homologous mammalian GlcNAc 2-epimerase, indicating that mammalian GlcNAc 2-epimerases also employ a deprotonation/reprotonation mechanism for epimerization. These biochemical and mechanistical findings in PhGn2E may help to gain a deeper understanding of the metabolism of GlcNAc and consequently sialic acids in living cells. Crystallization studies of the PhGn2E haloenzyme are a part of our current research program.

**Supplementary Materials:** The following are available online at www.mdpi.com/2073-4344/6/12/212/s1. Figure S1: Amino acids sequence alignment of GlcNAc 2-epimerases from *Bos taurus, Homo sapiens, Mus musculus, Sus scrofa, Anabaena* sp. CH1 and *Pedobacter heparinus*. Figures S2 and S3: NMR characterization of GlcNAc analogues. Figures S4–S6: MS analysis of the Neu5Acα2-6GalβX analogues. Figure S7: Thermo stability analysis of PhGn2E. Figure S8: The effect of detergents and denaturants on PhGn2E. Figure S9: Effects of nucleotides on PhGn2E. Figure S10: possible enzymatic epimerization mechanisms. Figure S11: Structural overlay of porcine Gn2E with modeled PhGn2E. Table S1: MALDI-TOF MS analysis of purified PhGn2E. Table S2: Primer sequences for generating mutant PhGn2E variants.

**Acknowledgments:** This work was supported in part by the Natural Science Foundation of China (grant numbers 31471703, A0201300537 and 31671854 to J.V. and L.L.), and the 100 Foreign Talents Plan (grant number JSB2014012 to J.V.). The authors would like to thank Louis Conway (GGBRC, Nanjing) for language editing of this manuscript.

**Author Contributions:** S.-Y.W. designed and performed experiments, analyzed the data and drafted the manuscript. P.L. designed NMR experiment, analyzed the NMR data and assisted in the manuscript preparation. A.-M.L. performed the NMR experiment. X.-C.D. cloned PhGn2E. H.-Y.M. performed mass spectrometric analysis. J.V. and L.L. conceived the project, designed experiments, and organized and finalized the manuscript.

**Conflicts of Interest:** The authors declare no conflict of interest.

## References

1. Inoue, S.; Kitajima, K. KDN (deaminated neuraminic acid): Dreamful past and exciting future of the newest member of the sialic acid family. *Glycoconj. J.* **2006**, *23*, 277–290. [CrossRef] [PubMed]
2. Schauer, R. Chemistry, metabolism, and biological functions of sialic acids. *Adv. Carbohydr. Chem. Biochem.* **1982**, *40*, 131–234. [PubMed]
3. Tzanakakis, G.N.; Syrokou, A.; Kanakis, I.; Karamanos, N.K. Determination and distribution of *N*-acetyl- and *N*-glycolylneuraminic acids in culture media and cell-associated glycoconjugates from human malignant mesothelioma and adenocarcinoma cells. *Biomed. Chromatogr.* **2006**, *20*, 434–439. [CrossRef] [PubMed]
4. Devine, P.L.; Clark, B.A.; Birrell, G.W.; Layton, G.T.; Ward, B.G.; Alewood, P.F.; McKenzie, I.F.C. The breast tumor-associated epitope defined by monoclonal-antibody 3E1.2 is an O-linked mucin carbohydrate containing *N*-glycolilneuraminic acid. *Cancer Res.* **1991**, *51*, 5826–5836. [PubMed]
5. Hedlund, M.; Tangvoranuntakul, P.; Takematsu, H.; Long, J.M.; Housley, G.D.; Kozutsumi, Y.; Suzuki, A.; Wynshaw-Boris, A.; Ryan, A.F.; Gallo, R.L.; et al. *N*-glycolylneuraminic acid deficiency in mice: Implications for human biology and evolution. *Mol. Cell. Biol.* **2007**, *27*, 4340–4346. [CrossRef] [PubMed]
6. Nasonkin, I.O.; Koliatsos, V.E. Nonhuman sialic acid Neu5Gc is very low in human embryonic stem cell-derived neural precursors differentiated with B27/N2 and noggin: Implications for transplantation. *Exp. Neurol.* **2006**, *201*, 525–529. [CrossRef] [PubMed]
7. Du, J.; Meledeo, M.A.; Wang, Z.Y.; Khanna, H.S.; Paruchuri, V.D.P.; Yarema, K.J. Metabolic glycoengineering: Sialic acid and beyond. *Glycobiology* **2009**, *19*, 1382–1401. [CrossRef] [PubMed]
8. Keppler, O.T.; Horstkorte, R.; Pawlita, M.; Schmidts, C.; Reutter, W. Biochemical engineering of the *N*-acyl side chain of sialic acid: Biological implications. *Glycobiology* **2001**, *11*, 11R–18R. [CrossRef] [PubMed]
9. Lawrence, S.M.; Huddleston, K.A.; Pitts, L.R.; Nguyen, N.; Lee, Y.C.; Vann, W.F.; Coleman, T.A.; Betenbaugh, M.J. Cloning and expression of the human *N*-acetylneuraminic acid phosphate synthase gene with 2-keto-3-deoxy-D-glycero-D-galacto-nononic acid biosynthetic ability. *J. Biol. Chem.* **2000**, *275*, 17869–17877. [CrossRef] [PubMed]
10. Stasche, R.; Hinderlich, S.; Weise, C.; Effertz, K.; Lucka, L.; Moormann, P.; Reutter, W. A bifunctional enzyme catalyzes the first two steps in *N*-acetylneuraminic acid biosynthesis of rat liver—Molecular cloning and functional expression of UDP-*N*-acetyl-glucosamine 2-epimerase/*N*-acetylmannosamine kinase. *J. Biol. Chem.* **1997**, *272*, 24319–24324. [CrossRef] [PubMed]
11. Blick, T.J.; Tiong, T.; Sahasrabudhe, A.; Varghese, J.N.; Colman, P.M.; Hart, G.J.; Bethell, R.C.; McKimmBreschkin, J.L. Generation and characterization of an influenza virus neuraminidase variant with decreased sensitivity to the neuraminidase-specific inhibitor 4-guanidino-Neu5Ac2en. *Virology* **1995**, *214*, 475–484. [CrossRef] [PubMed]
12. Lawrence, S.M.; Huddleston, K.A.; Tomiya, N.; Nguyen, N.; Lee, Y.C.; Vann, W.F.; Coleman, T.A.; Betenbaugh, M.J. Cloning and expression of human sialic acid pathway genes to generate CMP-sialic acids in insect cells. *Glycoconj. J.* **2001**, *18*, 205–213. [CrossRef] [PubMed]
13. Varki, A. Radioactive tracer techniques in the sequencing of glycoprotein oligosaccharides. *FASEB J.* **1991**, *5*, 226–235. [PubMed]
14. Luchansky, S.J.; Yarema, K.J.; Takahashi, S.; Bertozzi, C.R. GlcNAc 2-epimerase can serve a catabolic role in sialic acid metabolism. *J. Biol. Chem.* **2003**, *278*, 8035–8042. [CrossRef] [PubMed]
15. Ghosh, S.; Roseman, S. The sialic acids. V. *N*-acyl-D-glucosamine 2-epimerase. *J. Biol. Chem.* **1965**, *240*, 1531–1536. [PubMed]
16. Takahashi, S.; Hori, K.; Ogasawara, H.; Hiwatashi, K.; Sugiyama, T. Effects of Nucleotides on the interaction of renin with G1cNAc 2-epimerase (Renin binding protein, RnBP). *J. Biochem.* **2006**, *140*, 725–730. [CrossRef] [PubMed]

17. Takahashi, S.; Ogasawara, H.; Takahashi, K.; Hori, K.; Saito, K.; Mori, K. Identification of a domain conferring nucleotide binding to the *N*-acetyl-D-glucosamine 2-epimerase (renin binding protein). *J. Biochem.* **2002**, *131*, 605–610. [CrossRef] [PubMed]

18. Itoh, T.; Mikami, B.; Maru, I.; Ohta, Y.; Hashimoto, W.; Murata, K. Crystal structure of *N*-acyl-D-glucosamine 2-epimerase from porcine kidney at 2.0 angstrom resolution. *J. Mol. Biol.* **2000**, *303*, 733–744. [CrossRef] [PubMed]

19. Klermund, L.; Groher, A.; Castiglione, K. New *N*-acyl-D-glucosamine 2-epimerases from cyanobacteria with high activity in the absence of ATP and low inhibition by pyruvate. *J. Biotechnol.* **2013**, *168*, 256–263. [CrossRef] [PubMed]

20. Lee, Y.C.; Chien, H.C.R.; Hsu, W.H. Production of *N*-acetyl-D-neuraminic acid by recombinant whole cells expressing *Anabaena* sp. CH1N-acetyl-D-glucosamine 2-epimerase and *Escherichia coli* N-acetyl-D-neuraminic acid lyase. *J. Biotechnol.* **2007**, *129*, 453–460. [CrossRef] [PubMed]

21. Klermund, L.; Riederer, A.; Groher, A.; Castiglione, K. High-level soluble expression of a bacterial *N*-acyl-D-glucosamine 2-epimerase in recombinant *Escherichia coli*. *Protein Expr. Purif.* **2015**, *111*, 36–41. [CrossRef] [PubMed]

22. Sola-Carvajal, A.; Sanchez-Carron, G.; Garcia-Garcia, M.I.; Garcia-Carmona, F.; Sanchez-Ferrer, A. Properties of BoAGE2, a second *N*-acetyl-D-glucosamine 2-epimerase from *Bacteroides ovatus* ATCC 8483. *Biochimie* **2012**, *94*, 222–230. [CrossRef] [PubMed]

23. Liao, H.F.; Kao, C.H.; Lin, W.D.; Hsiao, N.W.; Hsu, W.H.; Lee, Y.C. *N*-Acetyl-D-glucosamine 2-epimerase from *Anabaena* sp. CH1 contains a novel ATP-binding site required for catalytic activity. *Process Biochem.* **2012**, *47*, 948–952. [CrossRef]

24. Samuel, J.; Tanner, M.E. Mechanistic aspects of enzymatic carbohydrate epimerization. *Nat. Prod. Rep.* **2002**, *19*, 261–277. [CrossRef] [PubMed]

25. Lee, Y.C.; Wu, H.M.; Chang, Y.N.; Wang, W.C.; Hsu, W.H. The central cavity from the (alpha/alpha)(6) barrel structure of *Anabaena* sp. CH1N-acetyl-D-glucosamine 2-epimerase contains two key histidine residues for reversible conversion. *J. Mol. Biol.* **2007**, *367*, 895–908. [CrossRef] [PubMed]

26. Ulrich, E.L.; Akutsu, H.; Doreleijers, J.F.; Harano, Y.; Ioannidis, Y.E.; Lin, J.; Livny, M.; Mading, S.; Maziuk, D.; Miller, Z. BioMagResBank. *Nucleic Acids Res.* **2008**, *36*, D402–D408. [CrossRef] [PubMed]

27. Tabata, K.; Koizumi, S.; Endo, T.; Ozaki, A. Production of *N*-acetyl-D-neuraminic acid by coupling bacteria expressing *N*-acetyl-D-glucosamine 2-epimerase and *N*-acetyl-D-neuraminic acid synthetase. *Enzym. Microb. Technol.* **2002**, *30*, 327–333. [CrossRef]

28. Moncla, B.; Braham, P.; Hillier, S. Sialidase (neuraminidase) activity among gram-negative anaerobic and capnophilic bacteria. *J. Clin. Microbiol.* **1990**, *28*, 422–425. [PubMed]

29. Maru, I.; Ohta, Y.; Murata, K.; Tsukada, Y. Molecular cloning and identification of *N*-acyl-D-glucosamine 2-epimerase from porcine kidney as a renin-binding protein. *J. Biol. Chem.* **1996**, *271*, 16294–16299. [CrossRef] [PubMed]

30. Takahashi, S.; Hori, K.; Takahashi, K.; Ogasawara, H.; Tomatsu, M.; Saito, K. Effects of nucleotides on *N*-acetyl-D-glucosamine 2-epimerases (renin-binding proteins): Comparative biochemical studies. *J. Biochem.* **2001**, *130*, 815–821. [CrossRef] [PubMed]

31. Chou, W.K.; Hinderlich, S.; Reutter, W.; Tanner, M.E. Sialic acid biosynthesis: Stereochemistry and mechanism of the reaction catalyzed by the mammalian UDP-*N*-acetylglucosamine 2-epimerase. *J. Am. Chem. Soc.* **2003**, *125*, 2455–2461. [CrossRef] [PubMed]

32. Wratil, P.R.; Rigol, S.; Solecka, B.; Kohla, G.; Kannicht, C.; Reutter, W.; Giannis, A.; Nguyen, L.D. A novel approach to decrease sialic acid expression in cells by a C-3-modified *N*-acetylmannosamine. *J. Biol. Chem.* **2014**, *289*, 32056–32063. [CrossRef] [PubMed]

33. Mahal, L.K.; Yarema, K.J.; Bertozzi, C.R. Engineering chemical reactivity on cell surfaces through oligosaccharide biosynthesis. *Science* **1997**, *276*, 1125–1128. [CrossRef] [PubMed]

34. Schwartz, E.L.; Hadfield, A.F.; Brown, A.E.; Sartorelli, A.C. Modification of sialic acid metabolism of murine erythroleukemia cells by analogs of *N*-acetylmannosamine. *Biochim. Biophys. Acta* **1983**, *762*, 489–497. [CrossRef]

35. Aich, U.; Meledeo, M.A.; Sampathkumar, S.G.; Fu, J.; Jones, M.B.; Weier, C.A.; Chung, S.Y.; Tang, B.C.; Yang, M.; Hanes, J.; et al. Development of delivery methods for carbohydrate-based drugs: Controlled release of biologically-active short chain fatty acid-hexosamine analogs. *Glycoconj. J.* **2010**, *27*, 445–459. [CrossRef] [PubMed]

36. Kim, E.J.; Sampathkumar, S.G.; Jones, M.B.; Rhee, J.K.; Baskaran, G.; Goon, S.; Yarema, K.J. Characterization of the metabolic flux and apoptotic effects of *O*-hydroxyl- and *N*-acyl-modified *N*-acetylmannosamine analogs in Jurkat cells. *J. Biol. Chem.* **2004**, *279*, 18342–18352. [CrossRef] [PubMed]

37. Laborda, P.; Wang, S.-Y.; Voglmeir, J. Influenza Neuraminidase Inhibitors: Synthetic Approaches, Derivatives and Biological Activity. *Molecules* **2016**, *21*, 1513. [CrossRef] [PubMed]

38. Cao, H.Z.; Li, Y.H.; Lau, K.; Muthana, S.; Yu, H.; Cheng, J.S.; Chokhawala, H.A.; Sugiarto, G.; Zhang, L.; Chen, X. Sialidase substrate specificity studies using chemoenzymatically synthesized sialosides containing C5-modified sialic acids. *Org. Biomol. Chem.* **2009**, *7*, 5137–5145. [CrossRef] [PubMed]

39. Yu, H.; Karpel, R.; Chen, X. Chemoenzymatic synthesis of CMP-sialic acid derivatives by a one-pot two-enzyme system: Comparison of substrate flexibility of three microbial CMP-sialic acid synthetases. *Bioorg. Med. Chem.* **2004**, *12*, 6427–6435. [CrossRef] [PubMed]

40. Jacobs, C.L.; Goon, S.; Yarema, K.J.; Hinderlich, S.; Hang, H.C.; Chai, D.H.; Bertozzi, C.R. Substrate specificity of the sialic acid biosynthetic pathway. *Biochemistry* **2001**, *40*, 12864–12874. [CrossRef] [PubMed]

41. Sampathkumar, S.G.; Li, A.V.; Yarema, K.J. Synthesis of non-natural ManNAc analogs for the expression of thiols on cell-surface sialic acids. *Nat. Protoc.* **2006**, *1*, 2377–2385. [CrossRef] [PubMed]

42. Ress, D.K.; Linhardt, R.J. Sialic acid donors: Chemical synthesis and glycosylation. *Curr. Org. Synth.* **2004**, *1*, 31–46. [CrossRef]

43. De Meo, C.; Priyadarshani, U. C-5 modifications in *N*-acetyl-neuraminic acid: Scope and limitations. *Carbohydr. Res.* **2008**, *343*, 1540–1552. [CrossRef] [PubMed]

44. Keppler, O.T.; Herrmann, M.; von der Lieth, C.W.; Stehling, P.; Reutter, W.; Pawlita, M. Elongation of the *N*-acyl side chain of sialic acids in MDCK II cells inhibits influenza A virus infection. *Biochem. Biophys. Res. Commun.* **1998**, *253*, 437–442. [CrossRef] [PubMed]

45. Lundgren, B.R.; Boddy, C.N. Sialic acid and *N*-acyl sialic acid analog production by fermentation of metabolically and genetically engineered *Escherichia coli*. *Org. Biomol. Chem.* **2007**, *5*, 1903–1909. [CrossRef] [PubMed]

46. Yu, H.; Chokhawala, H.A.; Huang, S.; Chen, X. One-pot three-enzyme chemoenzymatic approach to the synthesis of sialosides containing natural and non-natural functionalities. *Nat. Protoc.* **2006**, *1*, 2485–2492. [CrossRef] [PubMed]

47. Morgan, P.M.; Sala, R.F.; Tanner, M.E. Eliminations in the reactions catalyzed by UDP-*N*-acetylglucosamine 2-epimerase. *J. Am. Chem. Soc.* **1997**, *119*, 10269–10277. [CrossRef]

48. Mahuku, G.S. A simple extraction method suitable for PCR-based analysis of plant, fungal, and bacterial DNA. *Plant Mol. Biol. Rep.* **2004**, *22*, 71–81. [CrossRef]

49. Yu, H.; Huang, S.; Chokhawala, H.; Sun, M.; Zheng, H.; Chen, X. Highly efficient chemoenzymatic synthesis of naturally occurring and non-natural alpha-2,6-linked sialosides: A P. damsela alpha-2,6-sialyltransferase with extremely flexible donor-substrate specificity. *Angew. Chem.-Int. Ed.* **2006**, *45*, 3938–3944. [CrossRef] [PubMed]

50. Yao, H.L.; Conway, L.P.; Wang, M.M.; Huang, K.; Liu, L.; Voglmeir, J. Quantification of sialic acids in red meat by UPLC-FLD using indoxylsialosides as internal standards. *Glycoconj. J.* **2016**, *33*, 219–226. [CrossRef] [PubMed]

51. Laver, W.G. Purification, *N*-terminal amino acid analysis, and disruption of an influenza virus. *Virology* **1961**, *14*, 499–502. [CrossRef]

![catalysts logo] *catalysts*

MDPI

*Article*

# Photoassisted Oxidation of Sulfides Catalyzed by Artificial Metalloenzymes Using Water as an Oxygen Source †

Christian Herrero [1], Nhung Nguyen-Thi [2], Fabien Hammerer [1], Frédéric Banse [1], Donald Gagné [2], Nicolas Doucet [2], Jean-Pierre Mahy [1,*] and Rémy Ricoux [1,*]

[1]  Institut de Chimie Moléculaire et des Matériaux d'Orsay UMR 8182 CNRS, Bâtiment 420, Université Paris Sud, Université Paris Saclay, F-91405 Orsay CEDEX, France; christian.herrero@u-psud.fr (C.H.); fabien.hammerer@gmail.com (F.H.); frederic.banse@u-psud.fr (F.B.)
[2]  Institut National de la Recherche Scientifique (INRS)-Institut Armand-Frappier, Université du Québec, 531 Boulevard des Prairies, Laval, QC H7V 1B7, Canada; tepnhung@yahoo.com (N.N.-T.); donald.gagne@asrc.cuny.edu (D.G.); nicolas.doucet@iaf.inrs.ca (N.D.)
*  Correspondence: jean-pierre.mahy@u-psud.fr (J.-P.M.); remy.ricoux@u-psud.fr (R.R.); Tel.: +33-1-69-15-74-21 (J.-P.M.); +33-1-69-15-47-23 (R.R.)
†  Dedicated to the memory of Dr. Dominique Mandon.

Academic Editors: Jose M. Palomo and Cesar Mateo
Received: 19 October 2016; Accepted: 6 December 2016; Published: 12 December 2016

**Abstract:** The Mn(TpCPP)-Xln10A artificial metalloenzyme, obtained by non-covalent insertion of Mn(III)-meso-tetrakis(*p*-carboxyphenyl)porphyrin [Mn(TpCPP), **1-Mn**] into xylanase 10A from *Streptomyces lividans* (Xln10A) as a host protein, was found able to catalyze the selective photo-induced oxidation of organic substrates in the presence of $[Ru^{II}(bpy)_3]^{2+}$ as a photosensitizer and $[Co^{III}(NH_3)_5Cl]^{2+}$ as a sacrificial electron acceptor, using water as oxygen atom source.

**Keywords:** artificial metalloenzyme; manganese porphyrin; light induced oxidation

## 1. Introduction

Over the past few decades, the need for economically and environmentally compatible chemical processes has increased exponentially, along with global environmental concerns. As a result, the concept of green chemistry has emerged, which endorses the use of mild chemical reaction conditions, such as the use of non-polluting reactants, visible light as main energy source, and water as solvent, which, in addition, serves as O atom and electron sources [1]. In this context, oxidation reactions such as the conversion of alkanes into alcohols—one of first steps in the synthesis of products issued from the petrochemical industry—would benefit from significant improvements. Current chemical processes take place in wasteful organic solvents at high temperature, in addition to lacking in efficiency and/or selectivity [2,3]. Although promising metal catalysts have been developed and show encouraging implementation potential [4], several aspects such as efficacy, regio- and stereoselectivity, compound stability, or environmental toxicity remain to be addressed. In addition, these reactions often resort to oxidant reactants that might have a direct or indirect impact on the environment [2,3].

Natural enzymes provide us with astonishing examples of extremely efficient, selective, and stable catalysts. Oxidases constitute a widely investigated class of enzymes, most of which contain one or several metal ions at the heart of their redox center. For example, a non-heme iron enzyme such as phenylalanine hydroxylase (PAH) [5] catalyzes the hydroxylation of the aromatic side-chain of phenylalanine to generate tyrosine, whereas di-iron enzymes such as methane monooxygenases (MMO) allow the transformation of methane into methanol [6,7]. Cytochrome P450 enzymes constitute a large family of hemoproteins that can perform the oxidation of various substrates at high catalytic rates using molecular oxygen as the sole source of oxidant through a reductive activation mechanism [8]. During the catalytic cycle, two electrons are provided by redox mediators, Nicotinamide Adenine Dinucleotide (NADH), Nicotinamide Adenine Dinucleotide phosphate (NADPH), etc., allowing the formation of a very reactive high-valent iron-oxo intermediate responsible for substrate oxidation.

Over the years, P450 enzymes have been largely studied and utilized for catalytic purposes. However, electron delivery to the iron center remains the principal hurdle to high catalytic efficiency. On the one hand, natural redox cofactors are expensive, requiring the need to recycle them or to avoid their use. On the other hand, the electrons are not directly transferred to the heme iron, but instead are delivered through a P450 reductase that involves two flavin redox cofactors, i.e., Flavin Adenine Dinucleotide (FAD) and Flavin Mononucleotide (FMN). Two electrons are initially transferred from NADPH to FAD to yield $FADH_2$, which then transfers them to FMN. The obtained $FMNH_2$, then delivers sequentially, one by one, these two electrons to the iron atom at precise steps of the catalytic cycle [9]. Such a complex process, which is also driven by protein conformational changes, is then far from reproducible, and numerous approaches have been developed to mimic this catalytic reaction. For instance, approaches like the use of alternative oxygen atom donors (PhIO, ROOH, $H_2O_2$, $KHSO_5$, etc.) [10], chemical or electrochemical reductions [11], and, more recently, reductase proteins that were substituted with ruthenium-based photosensitizers capable of gathering electrons upon light irradiation and transferring them to cytochrome (P450) enzymes [12,13].

This latter field of research is appealing because it utilizes visible light as the sole energy input in the chemical reaction, which can be used either under reductive or oxidative experimental conditions. The former pathway relies on the photo-reduction by diethyldithiocarbamate (DTC) of a $Ru^{II}$-chromophore that is covalently attached to the apo-P450, which then catalyzes the reductive activation of oxygen leading, in the presence of $H^+$, to a high-valent $P450Fe^V$-oxo species that performs the oxidation of fatty acids [12,13]. Conversely, the oxidative pathway relies on the ability to quench the excited state of a $[Ru^{II}(bpy)_3]^{2+}$ chromophore with an irreversible electron acceptor such as $[Co^{III}(NH_3)_5Cl]^{2+}$, in order to form the highly oxidative $[Ru^{III}(bpy)_3]^{3+}$, which then catalyzes the two-electron oxidation of an iron-bound water molecule with formation of the high-valent iron-oxo species (Scheme 1, Path a) [14].

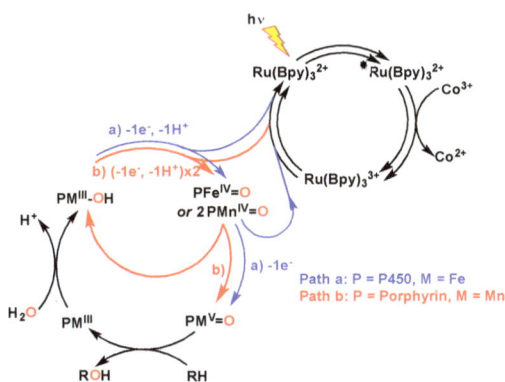

**Scheme 1.** Reaction pathways for the two-electron oxidation of an organic substrate by light-driven activation of a catalyst: Path (**a**) two-electron oxidation of an iron-bound water molecule into high-valent P-450-iron-oxo species by $[Ru^{III}(bpy)_3]^{3+}$, generated by quenching of the excited state of $[Ru^{II}(bpy)_3]^{2+}$ by an irreversible electron acceptor ($[Co^{III}(NH_3)_5Cl]^{2+}$) [14]; Path (**b**) one-electron oxidation of an $Mn^{III}$-OH complex into a porphyrin-$Mn^{IV}$-oxo species by $[Ru^{III}(bpy)_3]^{3+}$, followed by the disproportionation of the $Mn^{IV}$-oxo species into $Mn^{III}$ and $Mn^V$=O [15,16].

The oxidative pathway was also used with biomimetic systems involving manganese porphyrins as catalysts and water as an oxygen source, which were reported originally by Calvin's group [15], and more recently by Fukuzumi et al. [16]. In these systems, the authors showed that the photogenerated highly oxidative $[Ru^{III}(bpy)_3]^{3+}$ can oxidize the $Mn^{III}$-OH complex into an $Mn^{IV}$-oxo species. In this case, the $Mn^V$-oxo species was not directly formed by a second one-electron oxidation, but rather arose from the disproportionation of the $Mn^{IV}$-oxo species into $Mn^{III}$ and $Mn^V$=O (Scheme 1, Path b) [16]. Such a system could catalyze the oxidation of water-soluble substrates, including alkenes, alkanes, and sulfides [16]. Previous work from our group also showed that the moderately enantioselective light-induced oxidation of thioanisole by water as the oxygen atom source could occur in the presence of a system that associated a protein that induced chirality, BSA (Bovine Serum Albumin), a manganese corrole as oxygen atom transfer catalyst, and $[Ru^{II}(bpy)_3]^{2+}$ as photosensitizer [17].

In parallel, other work also allowed us to show the strong affinity of manganese sodium tetra-*para*-carboxylatophenyl-porphyrin (Mn-TCPP) for xylanase 10A (Xln10A) from *Streptomyces lividans*, a glycoside hydrolase that catalyzes the hydrolysis of $(1-4)$-β-D-xylosidic bonds in xylan biopolymers to release xylose [18]. Xylanase 10A is an abundant and resistant protein. It is easily purified and its sequence and three-dimensional structure are known, which makes it a good candidate as a protein scaffold for the production of biohybrid catalysts resulting from the combination of a natural protein and an artificial metal complex. We expected this combination of enzymatic and molecular catalysis to yield hybrid catalysts showing improved efficacy, selectivity, and stability. Studies on the reactivity of the Mn-TCPP-Xln10A hybrid showed that it possessed good mono-oxygenase activity for the epoxidation of styrene-type compounds, in addition to yielding important enantiomeric excesses, in particular for the epoxidation of *p*-methoxystyrene by oxone. This selectivity was further illustrated by molecular modeling studies that showed the existence of hydrogen bonds between the methoxy substituent of the substrate and a tyrosine residue of the protein [18].

We thus decided to use our Mn-TCPP-Xln10A biohybrid as a catalyst for the selective photo-induced oxidation of organic substrates in the presence of $[Ru^{II}(bpy)_3]^{2+}$ as a photosensitizer and $[Co^{III}(NH_3)_5Cl]^{2+}$ as a sacrificial oxidant, with water as an oxygen atom source. We report that both Mn-TCPP and its complex with xylanase 10A can catalyze the photoinduced sulfoxidation of thioanisole and 4-methoxythioanisole with slight enantiomeric excesses, using water as an oxygen atom source.

## 2. Results

### 2.1. Preparation of the Catalysts

Meso-tetrakis-*para*-carboxyphenylporphyrin **1** was first prepared in a 10.5% yield by condensation of p-carboxybenzaldehyde with pyrrole in propionic acid according to the procedure of Adler et al. [19,20]. Its Electon Spray Ionisation Mass (ESI-MS) and Ultra-Violet-visible (UV-vis) spectroscopy characteristics were found identical to those already reported [19]. The insertion of manganese was performed by treatment of **1** with an excess of $Mn(OAc)_2$, $4\ H_2O$ in 50 mM AcONa buffer, and pH 4.0 at 100 °C for 4 h, as described in the experimental section. The excess manganese salt was then removed by two successive purifications, first on P6DG exclusion gel, and second on Chelex. After lyophylization, **1**-Mn was obtained in a quantitative yield and characterized by ESI-MS and UV-vis spectroscopy, in agreement with those already reported [18] and with an Mn(T*p*CPP) structure (Figure 1).

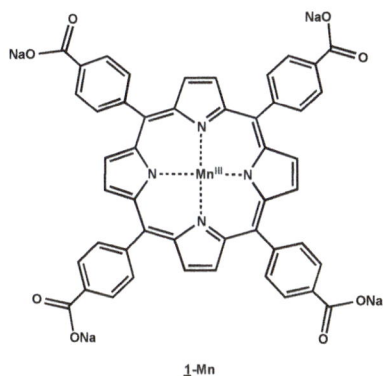

**Figure 1.** Manganese(III)-meso-tetrakis-*para*-carboxyphenylporphyrin, 1-Mn.

To prepare the **1**-Mn-Xln10A artificial hemoprotein, xylanase 10A (Xln10A) was first purified from the supernatant of *S. lividans* culture as reported earlier [21]. Solutions of the artificial hemoprotein were then prepared by incubating Xln10A in 50 mM sodium phosphate buffer, pH 7.0 with various amounts of **1**-Mn for 30 min. at room temperature. **1**-Mn-Xln10A was then characterized by UV-visible spectroscopy as described above, and its UV-visible spectrum was similar to that of **1**-Mn alone, with maxima at 468, 566 and 599 nm in 50 mM sodium phosphate buffer pH 7.0.

UV-visible spectroscopy could thus not be used for the determination of the stoichiometry and of the dissociation constant of the **1**-Mn-Xln10A. Since, in addition, **1**-Mn-Xln10A dissociates under mass spectrometry analysis conditions, the binding affinity of the **1**-Mn complex for Xln10A was then studied by measuring the quenching of the fluorescence of the tryptophan residues of Xln10A, 5 μM in 50 mM sodium phosphate buffer pH 7.0 at 25 °C as a function of the **1**-Mn concentration as previously reported [18]. The quenching was complete and selective, and suitable data were obtained from plots of fluorescence intensity versus the Xln10A/**1**-Mn ratio. A double reciprocal plot of the residual fluorescence intensity at 340 nm as a function of the **1**-Mn final concentration allowed calculation of a $K_D$ value of 1.5 μM for **1**-Mn-Xln10A and determined that only one ligand was bound per protein [18]. This $K_D$ value was similar to that already reported for the Fe(T*p*CPP)-Xln10A complex ($K_D$ = 0.5 μM) [21], and was in agreement with the fact that the binding of the porphyrin cofactor occurred thanks to interactions of three of its carboxylate substituents with amino-acid side chains of the protein and that no amino-acid side chain was interacting with the metal cation. This could be explained after molecular modeling calculations [18].

Taking into account the $K_D$ values measured for **1**-Mn-Xln10A, it was possible to calculate the theoretical concentration of protein necessary to bind 100% of the metal cofactor. It was thus found that four equivalents of Xln10A were necessary to bind one equivalent of the **1**-Mn cofactor. Accordingly, when the effect of the Xln10A/**1**-Mn ratio on the oxidation of styrene derivatives by KHSO₅ catalyzed by **1**-Mn-Xln10A in 50 mM phosphate buffer, pH 7.0, it appeared that the total yield of the reaction as well as its chemoselectivity and the enantiomeric excesses observed for the obtained epoxides were optimal when four equivalents of Xln10A were used with respect to **1**-Mn. It was thus decided to perform the photoassisted oxidation of sulfides in the presence of Xln10A and **1**-Mn in a 4/1 ratio as catalyst.

## 2.2. Catalysis

The oxidation of thioanisole (3200 eq.) 99 mM in 50 mM phosphate buffer solution (pH 7.0) was first assayed in the presence of 0.42 mM [Ru$^{II}$(bpy)₃]$^{2+}$ (13 eq.) as a photosensitizing agent, 12 mM [Co$^{III}$(NH₃)₅Cl]$^{2+}$ (390 eq.) as an irreversible electron acceptor and 32.5 μM **1**-Mn (1 eq.) in the presence of 130 μM Xln10A (4.2 eq.) (Xln10A/1-Mn ratio = 4) as catalyst (Scheme 2).

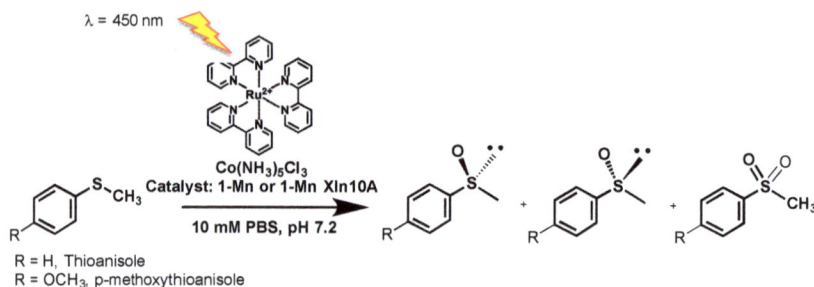

**Scheme 2.** Photoassisted oxidation of thioanisole and *p*-methoxythioanisole in the presence of $[Ru^{II}(bpy)_3]^{2+}$ as a photosensitizing agent, $[Co^{III}(NH_3)_5Cl]^{2+}$ as an irreversible electron acceptor and **1**-Mn or **1**-Mn-Xln10A as a catalyst.

When control experiments were run either in the absence of catalyst or in the absence of the $[Ru^{II}(bpy)_3]^{2+}$ photosensitizing agent, no oxidation products were formed. On the contrary, when the reaction was run in the presence of $[Ru^{II}(bpy)_3]^{2+}$ and of either **1**-Mn or **1**-Mn-Xln10A as catalyst, methylphenylsulfoxide was obtained as the major product. A Turnover number (TON) of $125 \pm 5$ was obtained after the photo driven reaction of **1**-Mn with $[Ru^{II}(bpy)_3]^{2+}$ in the absence of protein, which corresponds to a 64% yield with respect to $[Co^{III}(NH_3)_5Cl]^{2+}$. Minor amounts of sulfone ($6 \pm 1$ TON) were also formed (Table 1). Further HPLC analysis showed that a racemic mixture of methylphenylsulfoxide was produced, and thus that the reaction was not enantioselective. When the experiment was performed in the presence of Xln10A, methylphenylsulfoxide was obtained as the sole product with a TON of $25 \pm 2$ (13% yield with respect to $[Co^{III}(NH_3)_5Cl]^{2+}$). After HPLC analysis, no enantiomeric excess could be measured, showing that under those conditions the protein did not induce any stereoselectivity.

**Table 1.** Products obtained upon photoassisted oxidation of thioanisole in the presence of $[Ru^{II}(bpy)_3]^{2+}$ as a photosensitizing agent, $[Co^{III}(NH_3)_5Cl]^{2+}$ as electron acceptor, and various catalysts including **1**-Mn-Xln10A and **1**-Mn.

| Substrate | Catalyst | Products (TON) Sulfone | Sulfoxide | Enantiomeric Excess (ee) (%) | Reference |
|---|---|---|---|---|---|
| | **1**-Mn [a] | $6 \pm 1$ | $125 \pm 5$ | 0 | This work |
| | **1**-Mn-Xln10A [a] | - | $25 \pm 2$ | 0 | This work |
| Thioanisole ($R$ = H) | Mn-Corrole | 1.6 | $32 \pm 3$ | 0 | [17] |
| | Mn-Corrole-BSA | - | $21 \pm 7$ | 12–16 | [17] |
| | Ru$_{Phot}$-Ru$_{Cat}$-H$_2$O [b] | - | 745 | - | [22] |
| *p*-Methoxy-thioanisole | **1**-Mn [a] | - | $100 \pm 4$ | 0 | This work |
| ($R$ = OCH$_3$) | **1**-Mn-Xln10A [a] | - | $10 \pm 1$ | $7 \pm 1$ | This work |
| | Ru$_{Phot}$-Ru$_{Cat}$-H$_2$O [b] | - | 709 | - | [22] |
| *p*-Bromo-thioanisole | Ru$_{Phot}$-Ru$_{Cat}$-H$_2$O [b] | - | 314 | - | [22] |
| ($R$ = Br) | Ru$_{Phot}$-Ru$_{Cat}$-H$_2$O [c] | - | 201 | - | [23] |
| 2-(CH$_3$-thio)ethanol | Mn-TMPS [c] | | 30 | | [16] |

[a] Conditions as described in the experimental section; $[Ru^{II}(bpy)_3]^{2+}$ = ruthenium-tris-bipyrridyle, $[Co^{III}(NH_3)_5Cl]^{2+}$ = Cobalt-pentammine chloride, Xln 10A = xylanase 10A, Manganese(III)-meso-tetrakis-*para*-carboxyphenylporphyrin = **1**-Mn [b] Ru$_{Phot}$-Ru$_{Cat}$-H$_2$O = ([(bpy)$_2$Ru(4-Me-BPy-terPy)Ru(bpy)-(OH$_2$)]$^{4+}$; [b] Ru$_{Phot}$-Ru$_{Cat}$-H$_2$O = ([(bpy)$_2$Ru(tpphen)Ru(bpy)-(OH$_2$)]$^{4+}$, Cat/Substrate/$[Co^{III}(NH_3)_5Cl]^{2+}$ ratio = 1/500/1000; [c] Mn-TMPS (Mn(III)-5,10,15,20-tetrakis (2,4,6-trimethyl-3-sulfonatophenyl)porphyrin 0.1 mM, Cat/$[Ru^{III}(bpy)_3]^{2+}$/ Substrate/$[Co^{III}(NH_3)_5Cl]^{2+}$ ratio = 1/10/1000/1000.

We thus thought about investigating other sulfide derivatives that could lead to an enantioselective sulfoxidation under the same conditions. In this respect, we were inspired by the epoxidation of *p*-methoxy-styrene by oxone, in the presence of the Mn-TCPP-Xln10A hybrid as catalyst, which was previously demonstrated to be highly enantioselective (80% ee in favor of the *R* isomer). This was due to a preferential orientation of the *re* face of the styrene derivative towards the manganese-oxo oxidizing intermediate, provided by the formation of a H-bond between the O atom of the *p*-methoxy substituent and tyrosine 172 of Xln10A (Scheme 3A) [18]. We thus thought about using *p*-methoxy-thioanisole as substrate, expecting that in a similar way to *p*-methoxy-styrene, an H-bond between the O atom of the *p*-methoxy substituent and tyrosine 172 of Xln10A would preferentially orientate the sulfide towards the Mn-porphyrin-derived oxidizing species in order to get the R sulfoxide (Scheme 3B). The photoassisted sulfoxidation of 84 mM p-methoxy-thioanisole (2700 eq.) was then assayed in 50 mM phosphate buffer solution (pH 7.0) under the same conditions as those previously used for thioanisole.

**A)**   **B)**

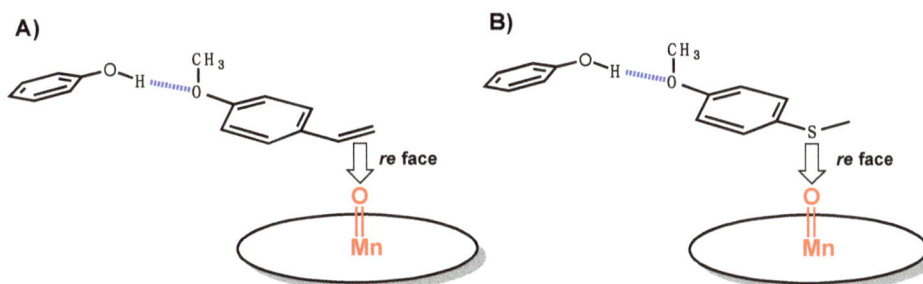

**Scheme 3.** Selective orientation of (**A**) *p*-methoxystyrene and (**B**) *p*-methoxythioanisole towards the putative porphyrin-Mn-oxo oxidizing species using **1**-Mn-Xln10A as catalyst

When **1**-Mn alone was used as an oxidizing catalyst, methyl-*p*-methoxyphenylsulfoxide was obtained as the only product with a TON of $100 \pm 4$ (51% yield with respect to $[Co^{III}(NH_3)_5Cl]^{2+}$), but the reaction was not stereoselective, as no enantiomeric excess could be measured after HPLC analysis. When the same reaction was performed in the presence of Xln10A, methyl-*p*-methoxyphenylsulfoxide was formed with a TON of $10 \pm 1$ (5% yield with respect to $[Co^{III}(NH_3)_5Cl]^{2+}$), but this time an enantiomeric excess of $7\% \pm 1\%$ in favor of the R Isomer could be measured, showing that under those conditions the protein induced a slight stereoselectivity in favor of the expected *R*-isomer.

### 2.3. Influence of the Xln10A/**1**-Mn Ratio on Catalysis

The influence of the Xln10A/**1**-Mn ratio on the stereo selectivity was investigated. For this purpose, the oxidation of 116 mM methoxy-thioanisole in 50 mM phosphate buffer solution (pH 7.0) was performed in the presence of 0.42 mM $[Ru^{II}(bpy)_3]^{2+}$ (14 eq.) and 12 mM $[Co^{III}(NH_3)_5Cl]^{2+}$ (390 eq.), as described above, using 37.0 μM **1**-Mn (1 eq.) as catalyst in the presence of increasing amounts of xylanase 10A, ranging from 37.0 μM (1 eq.) to 148 μM (4.2 eq.). After extraction and analysis of the products, it appeared that the TON decreased from 54 to 24 TON when the Xln10A/**1**-Mn ratio increased from 1 to 4.2, whereas, surprisingly, no change in the enantiomeric excess was observed. One likely explanation for this could be that, when the protein is in excess with respect to **1**-Mn, it acts as a competitive substrate for the electron withdrawing reaction by the highly oxidative $[Ru^{II}(bpy)_3]^{3+}$ complex, leading to protein radicals that would initiate protein oligomerization. Further experiments were then performed to verify this hypothesis using sodium dodecyl sulfate polyacrylamide gel electrophoresis (SDS-PAGE) analysis of the protein.

### 2.4. Evolution of the Catalyst during the Reaction

To explain the weak enantiomeric excesses observed during the light-induced oxidation of thioanisole derivatives, the xylanase 10A protein was examined by denaturing SDS-PAGE after the reaction. This experiment was performed with 130 μM Xln10A as catalyst in 50 mM phosphate buffer solution (pH 7.0), with 99 mM thioanisole, 0.42 mM $[Ru^{II}(bpy)_3]^{2+}$, 12 mM $[Co^{III}(NH_3)_5Cl]^{2+}$, and 32.5 μM **1-Mn** (Figure 2).

As shown in Figure 2, the starting pure Xln10A was characterized by a concentrated single band at 32 kDa, indicative of monomeric form, whereas, after the reaction, a band at higher MW (>250 kDa) could be observed, showing potential oligomerization of the protein over the course of the reaction.

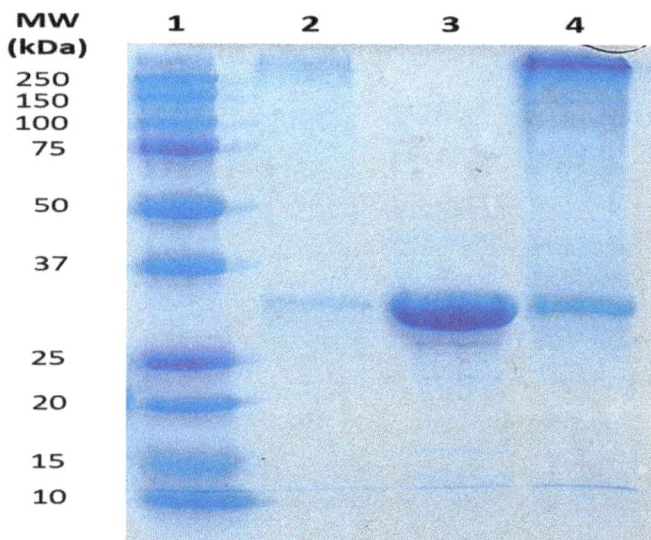

**Figure 2.** Sodium dodecyl sulfate polyacrylamide gel electrophoresis (SDS-PAGE) analysis of xylanase 10A. lane 1, molecular weight markers; lane 2, analysis of the pellet obtained after centrifugation of the reaction medium after the light induced oxidation reaction; lane 3, xylanase 10A purified before the reaction; lane 4, supernatant obtained after centrifugation of the reaction medium after catalysis.

*2.5. Mechanistic Studies*

Labeling experiments were realized to determine the origin of the oxygen atom of the obtained sulfoxide. Thus, the oxidation of *p*-methoxy-thioanisole was assayed under the same conditions as reported above, but 19.4% $H_2{}^{18}O$ was introduced for the preparation of the 50 mM phosphate buffer (pH 7.0) solvent. After extraction with ethyl acetate, the products were analyzed by High resolution (HR)-ESI MS. Twenty percent of the resulting sulfoxide contained $^{18}O$, as shown by the peaks observed at 171.0480 and 173.0477 that correspond to $[M + H]^+$ species for $CH_3S^{16}OC_6H_5$ and $CH_3S^{18}OC_6H_5$ methyl phenyl sulfoxides, respectively (Figure 3). This confirmed that the inserted O-atom in the oxidation product originated from water molecules of the solvent medium.

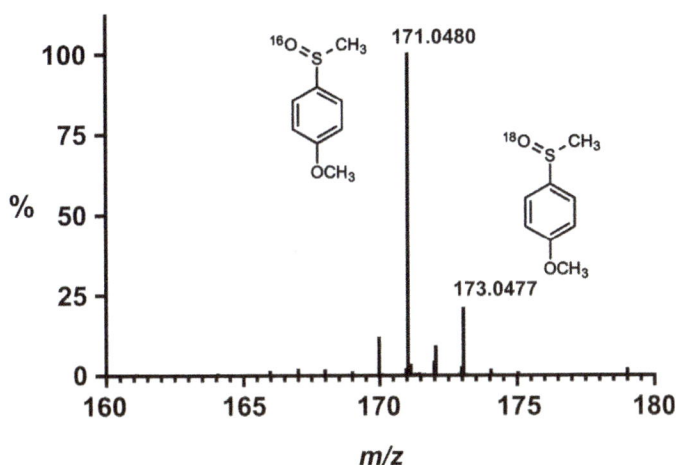

**Figure 3.** High-resolution Electron Spray Ionisation-Mass (ESI-MS) spectrum of the *p*-methoxy-thioanisole sulfoxide, obtained after the photoassisted oxidation of *p*-methoxy-thioanisole, in the presence of $[Ru^{II}(bpy)_3]^{2+}$, $[Co^{III}(NH_3)_5Cl]^{2+}$ and 1-Mn-Xln10A as catalyst, in 50 mM phosphate buffer (pH 7.0) containing 19% $H_2{}^{18}O$. Peaks at 171.0480 and 173.0477 correspond to $[M + H]^+$ species for $CH_3SOC_6H_5$ and $CH_3S^{18}OC_6H_5$ methyl phenyl sulfoxide, respectively.

## 3. Discussion

The aforementioned results show that both the manganese(III)-meso-tetrakis-*para*-carboxyphenylporphyrin 1-Mn and its complex with xylanase 10A, the 1-Mn-Xln 10A hemozyme, catalyze the photo-assisted oxidation of thioanisole derivatives with formation of the sulfoxide as major product (Scheme 2, Table 1). With both thioanisole and *p*-methoxy-thioanisole, the reaction is highly chemoselective and leads to the corresponding sulfoxide as a major product with greater yields with 1-Mn alone as catalyst (125 ± 5 TON and 100 ± 4 TON, respectively) than with 1-Mn-Xln10A as catalyst (25 ± 2 TON and 10 ± 1 TON, respectively). This could indicate that the insertion of **1**-Mn into the hydrophobic pocket of Xln10A [18] hinders the electron transfer from the manganese center to the photo-activated ruthenium complex. The results observed in the presence of 1-Mn-Xln10A are, however, comparable to those obtained under the same conditions for the oxidation of thioanisole catalyzed by an Mn-corrole complex and Mn-corrole-BSA artificial metalloenzyme, which respectively lead to the chemoselective formation of the corresponding sulfoxide with 32 ± 3 and 21 ± 7 TON (Table 1) [17]. In addition, as studies in the presence of $H_2{}^{18}O$ show the incorporation of $^{18}O$ in the obtained sulfoxide, and no reaction occurs in the absence of the $[Ru^{II}(bpy)_3]^{2+}$ photoactivator, it is reasonable to propose that the reaction follows a mechanism similar to that depicted in Scheme 1. Water would thus act as the oxygen source, leading to an $Mn^{III}$-OH that would be oxidized into an $Mn^{IV}$-oxo species by the photogenerated highly oxidative $[Ru^{III}(bpy)_3]^{3+}$ species. For the 1-Mn homogenous catalyst, the final $Mn^V$-oxo oxidizing species would likely arise from the disproportionation of the $Mn^{IV}$-oxo species into $Mn^{III}$ and $Mn^V$=O (Scheme 1, Path b) as reported for other manganese porphyrin complexes [15,16] whereas it would be formed upon oxidation of the $Mn^{IV}$-oxo by the photogenerated $[Ru^{III}(bpy)_3]^{3+}$ in the 1-Mn-Xln10A system (Scheme 1, pathway a) [14] because disproportionation between complexes embedded in Xln10A is unlikely.

Since the oxidation of 2-($CH_3$-thio)ethanol catalyzed by the Mn(III)-5,10,15,20-tetrakis (2,4,6-trimethyl-3-sulfonatophenyl)porphyrin (Mn-TMPS) under the same conditions was also reported to lead to the formation of the corresponding sulfoxide with 30 TON (Table 1) [16], 1-Mn-Xln10A is an equally interesting artificial metalloenzyme for the chemoselective photoinduced oxidation of sulfides in water. It is noteworthy, however, that these results illustrate lower yields relative to those obtained in the presence of supramolecular assemblies in which a ruthenium-tris-bipyridyl photoactivable moiety ($Ru_{Phot}$) is covalently attached to another $Ru(tpy)(bpy)(H_2O)$ catalytic moiety ($Ru_{cat}$) that photoinduces the sulfoxidation of thioanisole derivatives, such as thioanisole, *p*-Methoxy-, and *p*-bromo-thioanisole using water as an oxygen source, in the presence of $[Co^{III}(NH_3)_5Cl]^{2+}$ as sacrificial reductant with 200–745 TON (Table 1) [22,23].

As expected, no enantioselectivity was observed for the oxidation of thioanisole and *p*-methoxy thioanisole with that of **1-Mn** alone as catalyst (Table 1). Surprisingly, no enantioselectivity could be observed for the oxidation of thioanisole in the presence of Xln10A, which contrasted with the 12%–16% ee observed for the same reaction in the presence of Mn-Corrole-BSA as catalyst [17] (Table 1). The oxidation of *p*-methoxy-thioanisole, which was chosen in the hope of a selective pro-R positioning of the sulfide similar to that previously reported for *p*-methoxy-styrene [18] (Scheme 3) was only slightly enantioselective, though it led to the R sulfoxide as the major enantiomer, as was expected. However, the ee value, 7% ± 1%, was far from that obtained for the epoxidation of *p*-methoxy-styrene by oxone catalyzed by **1**-Mn-Xln10A, (80% ee in favor of the R epoxide). This could be due to a denaturation of the protein during catalysis. Indeed, when the Xln10A/**1-Mn** ratio increases, the ee remains constant, whereas the TON decreases, in agreement with a radical polymerization initiated by its direct reaction with the $[Ru^{II}(bpy)_3]^{3+}$ intermediate.

## 4. Materials and Methods

### 4.1. Physical Measurements

ESI-HRMS was determined on a micrOTOF-Q II 10027 apparatus (Bruker Daltonics, Billerica, MA, USA), UV/Vis spectra were recorded on an UVIKON-XL spectrophotometer (BioTek Instruments, Inc., Winooski, VE, USA) fluorescence spectra were recorded on a CARY Eclipse fluorometer (Agilent Technologies, Santa Clara, CA, USA). Gas chromatographs (Shimadzu Scientific Instruments, Inc., Columbia, MA, USA) were obtained with a Shimazu GC 2010 plus apparatus equipped with a Zebron ZB-SemiVolatiles Gas Chromatography (GC) column (30 m × 0.25 mm × 0.25 μm). HPLC (Agilent Technologies, Santa Clara, CA, USA) was performed on an Agilent Infinity 1260 apparatus equipped with a chiral column (Chiralcel OD-H, Daicel Co., Osaka, Japan, 250 mm × Φ4.6 mm).

### 4.2. Synthesis of Mn$^{III}$-meso-tetrakis(para-carboxyphenyl)porphyrin (1-Mn)

*meso*-Tetrakis(*para*-carboxyphenyl)porphyrin (**1**) was prepared in 10.5% yield by condensation of p-carboxybenzaldehyde with pyrrole in propionic acid according to the procedure of Adler et al. [19,20], and its molecular properties were found to be identical to those already reported [19]: $^1$H-NMR ($d_6$-DMSO) 8.85 (8H, s), 8.38 (8H, d, $J$ = 8 Hz), 8.32 (8H, d, $J$ = 8 Hz), −2.95 (2H, s, NH); Matrix-assisted Laser Desorption-Time Of Flight (MALDI-TOF)-MS: $m/z$ = 791.21 (M + H$^+$). The insertion of manganese was performed by treatment of **1** (2.5 mM) in AcONa buffer (pH 4.0, 50 mM) with $Mn(OAc)_2$, $4H_2O$ (11 equiv.) at 100 °C for 4 h. The progress of the reaction was monitored by UV/Vis spectroscopy. After 2 h under reflux, the Mn metalation was quantitative. The excess of manganese salt was then removed by successive purifications on P6DG exclusion gel and then on Chelex. After lyophylisation, **1-Mn** was obtained as a purple solid in a quantitative yield. Its properties were found to be identical to those already reported [18]: UV/Vis (pH 7.0, sodium phosphate buffer, 50 mM): $\lambda_{max}$ ($\varepsilon$): 468 (93 × 10$^3$ M$^{-1}$·cm$^{-1}$), 565, 600 nm; MALDI HR-MS (ESI): $m/z$ = 843.12396, calculated for $C_{48}H_{28}MnN_4O_8$, and $m/z$ = 843.12132 ($\Delta m$ = 5.04 ppm).

### 4.3. Preparation and Characterization of the Artificial Metalloenzyme 1-Mn-Xln10A

Xylanase 10A was first purified from the supernatant of *S. lividans* culture as reported earlier [19]. The various **1**-Mn-Xln10A hemozyme samples were then prepared by incubation for 30 min at room temperature (RT) of Xln10A in sodium phosphate buffer (50 mM, pH 7.0) with various amounts of

**1**-Mn. **1**-Mn-Xln10A was then characterized by UV/Vis and fluorescence spectroscopy studies as already reported [18].

*4.4. UV/Vis Spectroscopy Experiments*

**1**-Mn (5 µM) was incubated with Xln10A (1.5 equiv.) in 50 mM sodium phosphate buffer (pH 7.0) at room temperature for 30 min, and the UV/Vis spectrum was recorded between 280 nm and 700 nm. The spectrum obtained was found identical to that already reported [18]: UV/Vis (pH 7.0, sodium phosphate buffer, 50 mM): $\lambda_{max}$ ($\varepsilon$): 468 (92.3 × 10³ M⁻¹·cm⁻¹), 565,600 nm.

*4.5. Determination of the $K_D$ Values for **1**-Mn-Xln10A Complexes*

The $K_D$ value for **1**-Mn-Xln10A complexes as well as the **1**-Mn/Xln10A stoichiometry were determined as follows: the fluorescence spectrum of Xln10A, 5 µM in 50 mM sodium phosphate buffer, pH 7.2, at 25 °C was first recorded between 300 and 400 nm after excitation of the sample at 290 nm. The quenching of fluorescence was then followed by progressive addition of a 500 µM solution of **1**-Mn in 50 mM phosphate buffer pH 7.0. A double reciprocal plot of the residual fluorescence intensity at 340 nm as a function of the **1**-Mn final concentration afforded a $K_D$ value of 1.5 µM for the **1**-Mn-Xln10A complex and a 1/1 **1**-Mn/Xln10A stoichiometry.

*4.6. Photoassisted Oxidation of Sulfides*

For a typical photo-oxidation reaction, 150 µM of pure xylanase 10A, 37.5 µM porphyrin, 130 µL of 3.04 mM [Ru$^{II}$(bpy)$_3$]$^{2+}$ in phosphate buffer saline PBS, and either 11 µL thioanisole or 30 µL 4-methoxy thioanisole were added to a 800 µL degassed 50 mM phosphate buffer solution (pH 7.0) containing 14 mM [Co$^{III}$(NH$_3$)$_5$Cl]$^{2+}$. The solution was degassed by three freeze/thaw cycles, and the samples were illuminated with a 450 nm diode and stirred for 10 minutes at RT. The final concentrations of each component were 12 mM [Co$^{III}$(NH$_3$)$_5$Cl]$^{2+}$ (390 eq.), 130 µM xylanase 10A (4.2 eq.), 32.5 µM Mn-porphyrin (1 eq.), 0.42 mM [Ru$^{II}$(bpy)$_3$]$^{2+}$ (14 eq.), and 84 mM 4-methoxy thioanisole (2700 eq.), or 99 mM thioanisole (3200 eq.). Control experiments without protein were performed maintaining the above concentrations. Control experiments without chromophore were performed by substituting the [Ru$^{II}$(bpy)$_3$]$^{2+}$ solution by the equivalent PBS solution.

*4.7. Extraction and Analysis of the Oxidation Products*

After reaction, the content of the vial was eluted through a short (3 cm) silica plug to which 100 µL 10⁻² M acetophenone solution acting as internal standard had been previously added. The column was washed with 2 mL ethyl acetate in order to elute all the components. The aqueous phase of the resulting filtrate was removed and the organic phase was dried using NaSO$_4$ prior to GC analysis. For detection of thioanisole and its oxidation products, the applied temperature gradient was as follows: from 100 °C to 130 °C at 5 °C/min, and then 130 °C to 300 °C at 50 °C/min, which was further held constant for 3 min. The injector and flame injector detection (FID) temperature were set at 300 °C. Elution retention times were (min): acetophenone (4.03), thioanisole (4.32), thioanisole sulfoxide (7.56), and thioanisole sulfone (8.04). To study the enantiomeric excess of the obtained sulfoxide, the solvent was evaporated and the residue was re-dissolved in the minimum amount of isopropanol prior to HPLC analysis. The enantiomeric excesses were then determined by HPLC analysis using a chiral column (Chiralcel OD-H, Daicel Co., IlKirch, France, 250 mm × Φ4.6 mm). For the oxidation of thioanisole, samples were eluted with hexane/isopropanol (95:5, *v/v*, 1.0 mL/min), and, for the oxidation of *p*-methoxy-thioanisole, elution was done with hexane/isopropanol (97:3, *v/v*, 1.2 mL/min). All products were detected at 254 nm. Authentic samples of each sulfoxide enantiomer were prepared by the method previously described by Li et al. [24], and their retention time under the above described conditions are, respectively, 17 min and 21 min. for the *R* and *S* thioanisole sulfoxides, and 32 min and 36 min. for the *R* and *S* *p*-methoxy-thioanisole sulfoxides.

## 5. Conclusions

The aforementioned results demonstrate that an anionic manganese porphyrin, Mn(III)-meso-tetrakis-*para*-carboxyphenylporphyrin 1-Mn, as well as its complex with xylanase 10A, the 1-Mn-Xln10A artificial metalloenzyme, can be activated by visible light via a ruthenium chromophore. The resulting oxidized form of these complexes was found to catalyze the chemoselective and slightly enantioselective oxidation of thioanisole derivatives into the corresponding sulfoxides. The presence of $H_2^{18}O$ in the reaction medium led to the insertion of $^{18}O$ in the product, which confirmed the role of water as the O-atom source in the reaction. Thus, our results clearly illustrate that oxidation catalysis can be performed in aqueous medium, using no other oxygen atom source than water itself and using visible light as the only energy input. A slight enantiomeric excess in favor of the *R* sulfoxide could be obtained, which is likely to be related to protein polymerization and could be initiated by its direct reaction with intermediate ruthenium or Mn-porphyrin active species outside the binding site. A better strategy to limit those reactions, and then to increase the stereoselectivity of the reaction, could consist of covalently attaching the Ru complex to the protein, in a place close the catalyst, in order to minimize oxidation of the protein backbone, as it has already been realized in the case of cytochromes P450 [12,13].

**Acknowledgments:** This work was partly supported by the Programme Samuel-De-Champlain from the Commission Permanente de Coopération Franco-Québécoise (CPCFQ), by the LabEx de Chimie des Architectures Moléculaires Multifonctionnelles et des Matériaux (CHARM₃AT), by the ANR (project ANR Blanc Cathymetoxy (ANR-11-BS07-024) and by a Natural Sciences and Engineering Research Council of Canada (NSERC) Discovery Grant RGPIN 2016-05557 (to N.D.). D.G. held an NSERC Alexander Graham Bell Canada Graduate Scholarship.

**Author Contributions:** Christian Herrero and Rémy Ricoux conceived and designed the experiments; Donald Gagné, Fabien Hammerer, and Nhung Nguyen-Thi performed the experiments and contributed reagents/materials/analysis tools, Christian Herrero and Rémy Ricoux analyzed the data; Jean-Pierre Mahy, Frédéric Banse, Rémy Ricoux, and Nicolas Doucet wrote the paper.

**Conflicts of Interest:** The authors declare no conflict of interest.

## References

1. Li, C.J.; Anastas, P.T. Green Chemistry: present and future. *Chem. Soc. Rev.* **2012**, *41*, 1413–1414. [CrossRef] [PubMed]
2. Conley, B.L.; Tenn, W.J.; Young, K.J.H.; Ganesh, S.K.; Meier, S.K.; Ziatdinov, V.R.; Mironov, O.; Oxgaard, J.; Gonzales, J.; Goddard, W.A.; et al. Desig and study of homogeneous catalysts for the selective, low temperature oxidation of hydrocarbons. *J. Mol. Catal. A* **2006**, *251*, 8–23. [CrossRef]
3. Labinger, J.A. Selective alkane oxidation: Hot and cold approaches to a hot problem. *J. Mol. Catal. A* **2004**, *220*, 27–35. [CrossRef]
4. Collins, T.J. TAML oxidant activators: A new approach to the activation of hydrogen peroxide for environmentally significant problems. *Acc. Chem. Res.* **2002**, *35*, 782–790. [CrossRef] [PubMed]
5. Andersen, O.A.; Flatmark, T.; Hough, E. Crystal structure of the ternary complex of the catalytic domain of human phenylalanine hydroxylase with tetrahydrobiopterin and 3-(2-thienyl)-L-alanine and its implications for the mechanism of catalysis and substrate activation. *J. Mol. Biol.* **2002**, *320*, 1095–1108. [CrossRef]
6. Tinberg, C.E.; Lippard, S.J. Dioxygen Activation in Soluble Methane Monooxygenase. *Acc. Chem. Res.* **2011**, *44*, 280–288. [CrossRef] [PubMed]
7. Sirajuddin, S.; Rosenzweig, A.C. Enzymatic oxidation of methane. *Biochemistry* **2015**, *54*, 2283–2294. [CrossRef] [PubMed]
8. Hrycay, E.G.; Bandiera, S.M. Monooxygenase, peroxidase and peroxygenase properties and reaction mechanisms of cytochrome P450 enzymes. *Adv. Exp. Med. Biol.* **2015**, *851*, 1–61. [PubMed]
9. Iyanagi, T.; Xia, C.; Kim, J.J. NADPH-cytochrome P450 oxidoreductase: Prototypic member of the diflavin reductase family. *Arch. Biochem. Biophys.* **2012**, *528*, 72–89. [CrossRef] [PubMed]
10. Meunier, B.; Robert, A.; Pratviel, G.; Bernadou, J. Metalloporphyrins in catalytic oxidations and oxidative DNA cleavage. In *The Porphyrin Handbook*; Kadish, K.M., Smith, K., Guilard, R., Eds.; Academic Press: San Diego, FL, USA, 2000; Volume 4, pp. 119–184.

11. Meunier, B. Metalloporphyrins as versatile catalysts for oxidation reactions and oxidative DNA cleavage. *Chem. Rev.* **1992**, *92*, 1411–1456. [CrossRef]

12. Tran, N.H.; Nguyen, D.; Dwaraknath, S.; Mahadevan, S.; Chavez, G.; Nguyen, A.; Dao, T.; Mullen, S.; Nguyen, T.A.; Cheruzel, L.E. An Efficient Light-Driven P450 BM3 Biocatalyst. *J. Am. Chem. Soc.* **2013**, *135*, 14484–14487. [CrossRef] [PubMed]

13. Ener, M.E.; Leeb, Y.-T.; Winkler, J.R.; Gray, H.B.; Cheruzel, L.E. Photooxidation of cytochrome P450-BM3. *Proc. Natl. Acad. Sci. USA* **2010**, *107*, 18783–18786. [CrossRef] [PubMed]

14. Kato, M.; Nguyen, D.; Gonzalez, M.; Cortez, A.; Mullen, S.E.; Cheruzel, L.E. Regio- and stereoselective hydroxylation of 10-undecenoic acid with a light-driven P450 BM3 biocatalyst yielding a valuable synthon for natural product synthesis. *Bioorg. Med. Chem.* **2014**, *22*, 5687–5691. [CrossRef] [PubMed]

15. Maliyackel, A.C.; Otvos, J.W.; Spreer, L.O.; Calvin, M. Photoinduced oxidation of a water-soluble manganese(III) porphyrin. *Proc. Natl. Acad. Sci. USA* **1986**, *83*, 3572–3574. [CrossRef] [PubMed]

16. Fukuzumi, S.; Kishi, T.; Kotani, H.; Lee, Y.-M.; Nam, W. Highly efficient photocatalytic oxygenation reactions using water as an oxygen source. *Nat. Chem.* **2011**, *3*, 38–41. [CrossRef] [PubMed]

17. Herrero, C.; Quaranta, A.; Ricoux, R.; Trehoux, A.; Mahammed, A.; Gross, Z.; Banse, F.; Mahy, J.P. Oxidation catalysis via visible-light water activation of a [Ru(bpy)$_3$]$^{2+}$ chromophore BSA–metallocorrole couple. *Dalton Trans.* **2016**, *45*, 706–710. [CrossRef] [PubMed]

18. Allard, M.; Dupont, C.; Munoz Robles, V.; Doucet, N.; Lledos, A.; Marechal, J.-D.; Urvoas, A.; Mahy, J.-P.; Ricoux, R. Incorporation of manganese complexes into xylanase: New artificial metalloenzymes for enantioselective epoxidation. *ChemBioChem* **2012**, *13*, 240–251. [CrossRef] [PubMed]

19. Ricoux, R.; Allard, M.; Dubuc, R.; Dupont, C.; Marechal, J.-D.; Mahy, J.-P. Selective oxidation of aromatic sulfide catalyzed by an artificial metalloenzyme: New activity of hemozymes. *Org. Biomol. Chem.* **2009**, *7*, 3208–3211. [CrossRef] [PubMed]

20. Adler, A.D.; Longo, F.R.; Finarelli, J.D.; Goldmacher, J.; Assour, J.; Korsakoff, L.J. A simplified synthesis for meso-tetraphenylporphine. *J. Org. Chem.* **1967**, *32*, 476. [CrossRef]

21. Ricoux, R.; Dubuc, R.; Dupont, C.; Marechal, J.D.; Martin, A.; Sellier, M.; Mahy, J.P. Hemozymes peroxidase activity of artificial hemoproteins constructed from the *Streptomyces lividans* xylanase a and Iron(III)-carboxy-substituted porphyrins. *Bioconj. Chem.* **2008**, *19*, 899–910. [CrossRef] [PubMed]

22. Li, T.-T.; Li, F.-M.; Zhao, W.-L.; Tian, Y.-H.; Chen, Y.; Cai, R.; Fu, W.-F. Highly efficient and selective photocatalytic oxidation of sulfide by a chromophore-catalyst dyad of ruthenium-based complexes. *Inorg. Chem.* **2015**, *54*, 183–191. [CrossRef] [PubMed]

23. Hamelin, O.; Guillo, P.; Loiseau, F.; Boissonnet, M.-F.; Menage, S. A Dyad as photocatalyst for light-driven sulfide oxygenation with water as the unique oxygen atom source. *Inorg. Chem.* **2011**, *50*, 7952–7954. [CrossRef] [PubMed]

24. Li, A.T.; Zhang, J.D.; Xu, J.H.; Lu, W.Y.; Lin, G.Q. Isolation of *Rhodococcus* sp. ECU0066, a new sulfide monooxygenase producing strain for asymmetric sulfoxidation. *Appl. Environ. Microbiol.* **2009**, *75*, 551–556. [CrossRef] [PubMed]

*catalysts*

MDPI

*Article*

# Mechanistic and Structural Insight to an Evolved Benzoylformate Decarboxylase with Enhanced Pyruvate Decarboxylase Activity

**Forest H. Andrews [1], Cindy Wechsler [2], Megan P. Rogers [1], Danilo Meyer [2], Kai Tittmann [2] and Michael J. McLeish [1,\*]**

[1]  Department of Chemistry and Chemical Biology, Indiana University-Purdue University Indianapolis, Indianapolis, IN 46202, USA; fandrewsdpu@gmail.com (F.H.A.); megahugh@iupui.edu (M.P.R.)
[2]  Göttingen Center for Molecular Biosciences, Georg-August-University Göttingen, Ernst-Caspari-Haus, Justus-von-Liebig-Weg 11, 37077 Göttingen, Germany; wechsler.cindy@gmail.com (C.W.); dmeyer@ipb-halle.de (D.M.); Kai.Tittmann@biologie.uni-goettingen.de (K.T.)
\*  Correspondence: mcleish@iupui.edu; Tel.: +1-317-274-6889

Academic Editors: Jose M. Palomo and Cesar Mateo
Received: 1 November 2016; Accepted: 28 November 2016; Published: 30 November 2016

**Abstract:** Benzoylformate decarboxylase (BFDC) and pyruvate decarboxylase (PDC) are thiamin diphosphate-dependent enzymes that share some structural and mechanistic similarities. Both enzymes catalyze the nonoxidative decarboxylation of 2-keto acids, yet differ considerably in their substrate specificity. In particular, the BFDC from *P. putida* exhibits very limited activity with pyruvate, whereas the PDCs from *S. cerevisiae* or from *Z. mobilis* show virtually no activity with benzoylformate (phenylglyoxylate). Previously, saturation mutagenesis was used to generate the BFDC T377L/A460Y variant, which exhibited a greater than 10,000-fold increase in pyruvate/benzoylformate substrate utilization ratio compared to that of *wt*BFDC. Much of this change could be attributed to an improvement in the $K_m$ value for pyruvate and, concomitantly, a decrease in the $k_{cat}$ value for benzoylformate. However, the steady-state data did not provide any details about changes in individual catalytic steps. To gain insight into the changes in conversion rates of pyruvate and benzoylformate to acetaldehyde and benzaldehyde, respectively, by the BFDC T377L/A460Y variant, reaction intermediates of both substrates were analyzed by NMR and microscopic rate constants for the elementary catalytic steps were calculated. Herein we also report the high resolution X-ray structure of the BFDC T377L/A460Y variant, which provides context for the observed changes in substrate specificity.

**Keywords:** Thiamin diphosphate; X-ray crystallography; NMR spectroscopy; enzyme evolution

## 1. Introduction

The decarboxylases form the largest group of thiamin diphosphate (ThDP)-dependent enzymes [1]. The archetypal member of this group is pyruvate decarboxylase (PDC), which catalyzes the nonoxidative decarboxylation of pyruvate to yield acetaldehyde and carbon dioxide, a reaction critical to the fermentation pathway of several yeast and bacteria [2]. X-ray structures of PDCs from a variety of species show that, in addition to ThDP, the active site contains two ionizable acidic residues as well as two contiguous histidine residues that are located on an ordered loop [3–6]. The latter has been termed the HH-motif, and mutagenesis and kinetic studies have revealed that both histidines play significant roles in catalysis [7–10]. X-ray structures of several other ThDP-dependent decarboxylases reveal that, even though a variety of residues may line the substrate-binding pocket, thereby explaining the observed differences in substrate specificity, most possess the HH-motif and the two acidic residues [11–13].

Benzoylformate decarboxylase (BFDC) is the penultimate enzyme in the mandelate pathway, a secondary metabolic pathway that allows various pseudomonads to grow using *R*-mandelate as their only source of carbon [14–16]. Also a ThDP-dependent enzyme, BFDC catalyzes the nonoxidative decarboxylation of benzoylformate, generating benzaldehyde and carbon dioxide [14]. Given the similarity of their reactions, and their using the same cofactor, it may have been expected that the active sites of PDC and BFDC would be similar. However, this proved not to be the case, for the active site of BFDC was found to contain two leucine residues occupying positions equivalent to the HH-motif [17]. That said, the active site of BFDC does contain two histidine residues (His70 and His281) and mutagenesis studies suggest that they too are catalytically important [18,19]. Another difference was that the BFDC active site lacked the two acidic residues found in the HH-motif family. Instead, BFDC's only other ionizable residue is a serine, Ser26, which has also been implicated in the catalytic mechanism [18].

Although their active sites and substrate preference may differ, the catalytic cycles of BFDC and PDC can still be broken into the same series of steps (Scheme 1) [20]. The first two steps are the deprotonation of the C2 carbon of ThDP to generate either an ylid [16,21,22] or a carbene [23], followed by non-covalent binding of the 2-keto substrate to provide the Michaelis complex. Addition of the C2 carbanion/carbene of ThDP to the carbonyl of the substrate gives rise to the first covalent tetrahedral intermediate, which subsequently loses carbon dioxide resulting in the formation of carbanion/enamine intermediate typical of all ThDP-dependent enzymes. In decarboxylases such as BFDC and PDC, protonation of this enamine generates a second tetrahedral intermediate, which finally breaks down, releasing the product aldehyde and regenerating ThDP.

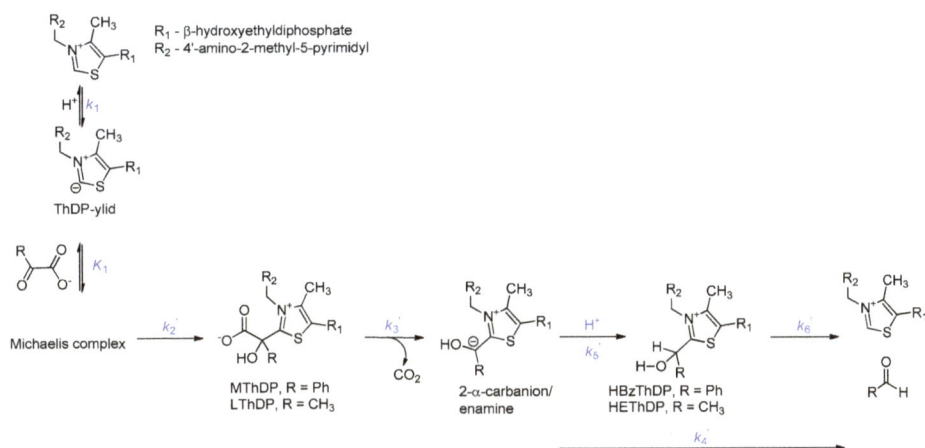

**Scheme 1.** Mechanism of benzoylformate decarboxylase (BFDC) and pyruvate decarboxylase (PDC) with intermediates and net rate constants.

Over the past few years, the steady-state distribution of the covalent intermediates has been used to determine net rate constants for the elementary catalytic steps of several ThDP-dependent enzymes including PDC [24] and BFDC [25]. For the former, decarboxylation and product release were found to be partially rate-limiting, but formation of lactyl-ThDP (LThDP), the first covalent intermediate, was quite rapid [24]. Conversely, the rate-limiting step for BFDC was found to be the formation of the first covalent intermediate, mandelyl-ThDP (MThDP) [18,25]. Further, BFDC was found to decarboxylate its first covalent adduct an order of magnitude faster than any other ThDP-dependent enzyme studied to date [25].

As with many ThDP-dependent decarboxylases, BFDC and PDC are both are capable of stereoselective carboligation reactions leading to chiral α-hydroxy ketones [26]. These have been used as versatile building blocks in the pharmaceutical (and other) industries [26–28]. As a consequence there is considerable interest in expanding the capabilities of the decarboxylases to accept alternative substrates. While both BFDC and PDC exhibit some substrate promiscuity neither enzyme is particularly efficient at decarboxylating the other's *in vivo* substrate [29]. Recently, as part of our ongoing efforts to understand the evolution of substrate specificity in ThDP-dependent enzymes, we used site-saturation mutagenesis in an attempt to convert a BFDC into a PDC [30]. The BFDC T377L/A460Y variant was observed during screening to have greatly enhanced PDC activity. Subsequently it was shown to have a 10,000-fold increase in its substrate utilization ratio between pyruvate and benzoylformate compared to that of *wt*BFDC [30].

Here we describe our efforts to identify the reasons for that change in specificity. In the first instance, we used NMR spectroscopy to analyze the intermediates obtained in the reaction of BFDC T377L/A460Y with both benzoylformate and pyruvate. The microscopic rate constants were calculated for the elementary catalytic steps in the reaction of both substrates. Further, in an attempt to understand the structural basis for the altered substrate specificity exhibited by the T377L/A460Y variant, its high resolution X-ray structure was determined.

## 2. Results and Discussion

### 2.1. Substrate Profile of T377L/A460Y

Our earlier study compared the reactions of *wt*BFDC, *Sc*PDC and the T377L/A460Y variant only with benzoylformate and pyruvate. Here the comparison was extended to include a variety of straight- and branched-chain 2-keto acids, as well as phenylpyruvate (Table 1). The steady-state kinetic constants for reaction of *wt*BFDC with pyruvate could only be determined under V/K conditions, i.e., at substrate concentrations well below the $K_m$ value, which was estimated to be easily in excess of 100 mM [30]. By contrast the T377L/A460Y variant has a $K_m$ value for pyruvate of 4.6 mM, which is comparable to the pyruvate $S_{0.5}$ value for *Sc*PDC. In addition to improving utilization of pyruvate, the double variant was found to decarboxylate several other aliphatic 2-keto acids at a faster rate than *wt*BFDC. While it was most efficient with 2-ketohexanoate and 4-methylthio-2-ketobutanoate as substrates, there was only a 60-fold difference in $k_{cat}/K_m$ values between the best substrate, 2-ketohexanoate, and its least efficient substrate, pyruvate. In many ways pyruvate is still the exception as its $K_m$ value was significantly higher than those of other aliphatic substrates. With regards to $k_{cat}$, there is less than a 15-fold decrease between the substrate that is turned over the fastest, 3-methyl-2-ketopentanoate, and the substrate that the T377L/A460Y variant decarboxylates the slowest, 2-ketopentanoate. With the exception of benzoylformate, the T377L/A460Y variant was able to decarboxylate its substrates more efficiently than *wt*BFDC. Similarly, only pyruvate and its longer chain analogues, 2-ketobutanoate and 2-ketopentanoate were better substrates for *Sc*PDC.

Overall, the data in Table 1 clearly demonstrate that the T377L/A460Y variant is able to decarboxylate a wide range of aliphatic substrates and, in most cases, more efficiently that either *wt*BFDC or *Sc*PDC. That said, it is also apparent that the ability of this variant to decarboxylate either benzoylformate or pyruvate is considerably less than that of *wt*BFDC or *Sc*PDC, respectively. The similarity of substrate $K_m$ values suggests that substrate binding is not likely to be the issue, rather the changes in active site geometry are having a significant effect on one or more of the individual catalytic steps.

**Table 1.** Kinetic characterization of *wt*BFDC, BFDC T377L/A460Y and ScPDC at pH 6.0 [a].

| Substrate | *wt*BFDC | | | BFDC T377L/A460Y | | | ScPDC | | |
|---|---|---|---|---|---|---|---|---|---|
| | $K_m$ (mM) | $k_{cat}$ (s$^{-1}$) | $k_{cat}/K_m$ (mM$^{-1}$·s$^{-1}$) | $K_m$ (mM) | $k_{cat}$ (s$^{-1}$) | $k_{cat}/K_m$ (mM$^{-1}$·s$^{-1}$) | $S_{0.5}$ (mM) | $k_{cat}$ (s$^{-1}$) | $k_{cat}/S_{0.5}$ (mM$^{-1}$·s$^{-1}$) |
| Benzoylformate | 0.27 ± 0.02 | 320 ± 4 | 1180 | 0.45 ± 0.04 | 15 ± 1 | 34 | NAD [b] | NAD [b] | NAD [b] |
| Pyruvate | - | - | 0.01 | 4.6 ± 0.3 | 3.3 ± 0.1 | 0.72 | 1.3 ± 0.1 | 60 ± 2 | 45 |
| 4-Methylthio-2-ketobutanoate | 1.3 ± 0.2 | 4.5 ± 0.3 | 3.5 | 0.14 ± 0.01 | 4.6 ± 0.1 | 33 | 0.80 ± 0.05 | 5.2 ± 0.3 | 6.6 |
| 2-Ketohexanoate | 5.3 ± 0.3 | 4.1 ± 0.1 | 1.3 | 0.14 ± 0.01 | 5.9 ± 0.2 | 42 | 1.5 ± 0.1 | 15 ± 1 | 9.8 |
| 2-Ketopentanoate | 11 ± 1 | 6.0 ± 0.5 | 1.8 | 0.15 ± 0.01 | 1.1 ± 0.1 | 7.5 | 1.1 ± 0.1 | 27 ± 2 | 24 |
| 4-Methyl-2-ketopentanoate | 9.6 [c] | 11 [c] | 1.1 [c] | 0.14 ± 0.02 | 2.7 ± 0.2 | 20 | 5.0 ± 0.2 | 10 ± 1 | 2.0 |
| 3-Methyl-2-ketopentanoate | - | - | 1.6 [b] | 1.2 ± 0.1 | 16 ± 1 | 13 | 8.7 ± 0.5 | 22 ± 1 | 2.5 |
| 2-Ketobutanoate | 4.1 ± 0.1 | 7.5 ± 0.6 | 0.5 | 0.39 ± 0.02 | 6.0 ± 0.1 | 16 | 0.49 ± 0.02 | 25 ± 1 | 49 |
| 3-Methyl-2-ketobutanoate | - | - | 2.6 [c,d] | 0.77 ± 0.06 | 12 ± 1 | 15 | 0.73 ± 0.03 | 6.3 ± 0.2 | 8.6 |
| Phenylpyruvate | - | - | 0.1 [c,d] | 0.82 ± 0.06 | 2.0 ± 0.1 | 2.5 | 0.23 ± 0.01 | 0.87 ± 0.02 | 3.7 |

[a] Reactions were carried out in 100 mM KPO$_4$ buffer containing 1 mM MgSO$_4$ and 0.5 mM ThDP. Values are the mean of three independent determinations ± standard error; [b] NAD, no activity detected; [c] Data from Siegert et al. [29]; [d] Specific activity determined at a substrate concentration of 20 mM.

## 2.2. Intermediate Distribution Analysis for Reaction of T377L/A460Y with Benzoylformate

To explore this possibility the microscopic rate constants for the BFDC T377L/A460Y variant were determined for the decarboxylation of both benzoylformate and pyruvate. These have been determined previously for the reaction of *wt*BFDC with benzoylformate [25]. In that study it was shown that the rate-limiting step was the formation of MThDP by addition of the ThDP ylid to benzoylformate, and that BFDC carried out its decarboxylation step at least two orders of magnitude faster than any other ThDP-dependent enzyme [25].

All three major intermediates, unreacted ThDP, MThDP, and 2-hydroxybenzyl-ThDP (HBzThDP), can be detected at steady-state when T377L/A460Y reacts with benzoylformate (Figure 1). While the MThDP signal is close to the limits of resolution, it is clearly visible. Following correction for non-enzymatic decarboxylation of MThDP [25] analysis reveals that ~37% of the active sites contain C2-unsubstituted ThDP and 3% contain MThDP. The remaining 60% are occupied by HBzThDP. This is distinctly different to the result obtained with *wt*BFDC where a similar analysis provided values of 79%, 3% and 18% for ThDP, MThDP and HBzThDP, respectively [25]. A comparison of the derived kinetic constants for wtBFDC and T377L/A460Y can be seen in Table 2. In *wt*BFDC addition of the ThDP ylid to benzoylformate (i.e., C–C formation, $k_2'$) is clearly rate limiting. Conversely, protonation of the enamine/elimination of benzaldehyde ($k_4'$) has been reduced ~100-fold in the T377L/A460Y variant and has now become rate limiting. It is notable that for both variants decarboxylation of MThDP ($k_3'$) is significantly faster than other steps in the reaction (Table 2). Further, even though the rate for T377L/A460Y has decreased considerably compared to that of the wild-type enzyme it is still faster than decarboxylation of LThDP by pyruvate-utilizing enzymes [20,31,32].

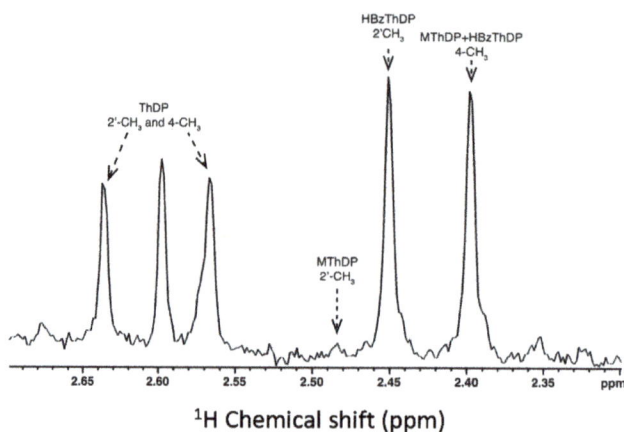

**Figure 1.** The $2'$-CH$_3$ and 4-CH$_3$ singlet signals of thiamin diphosphate (ThDP) and of benzoylformate-conjugated covalent intermediates were used for the assignment and quantitative analysis of the following intermediates: ThDP (2.64 and 2.57 ppm), mandelyl-ThDP (MThDP) (2.48 and 2.41 ppm), and 2-hydroxybenzyl-ThDP (HBzThDP) (2.45 and 2.40 ppm). Intermediates and cofactor were isolated from the protein by acid quench after a reaction time of 1 s and analyzed by $^1$H NMR at pH 0.75 at 25 °C as described in Materials and Methods. Relative concentrations of intermediates were determined from the relative integrals of the corresponding signals, which were used to calculate all microscopic rate constants of the catalytic cycle.

## 2.3. Intermediate Distribution Analysis for Reaction of T377L/A460Y with Pyruvate

The microscopic rate constants for reaction of *wt*BFDC with pyruvate could not be determined as only trace amounts of the reaction intermediates were observable at steady-state using the NMR-based approach. This is not surprising as activity of *wt*BFDC with pyruvate could only be measured under *V/K* conditions [30]. The lack of any observable reaction intermediates, coupled with a pyruvate $K_m$ value in excess of 100 mM, indicates that the rate-limiting step is the initial formation of the Michaelis complex and/or the subsequent C–C bond formation ($k_2'$) that provides LThDP.

For T377LA460Y the steady-state distribution of intermediates showed that, when pyruvate is used as substrate (Figure 2), all the key catalytic intermediates are present: ThDP, 2-lactyl-ThDP (LThDP) and 2-hydroxyethyl-ThDP (HEThDP). This finding indicates that all major steps of catalysis are on the same time scale. In addition to those anticipated intermediates, a minor fraction of 2-acetolactate-ThDP (ALThDP) was observed [31].

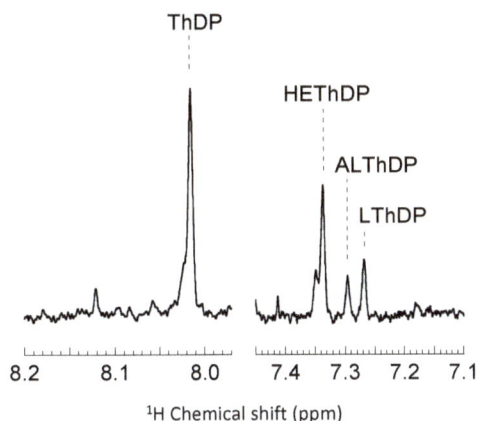

**Figure 2.** The C6′-H singlet signals of ThDP (thiamin diphosphate) and of pyruvate-conjugated covalent intermediates were used for the assignment and quantitative analysis of the following intermediates: ThDP (8.01 ppm), LThDP (lactyl-ThDP) (7.27 ppm), HEThDP (2-hydroxyethyl-ThDP) (7.34 ppm) and ALThDP (2-acetolactate-ThDP) (7.30 ppm). Intermediates and cofactor were isolated by acid quench after a reaction time of 8 s and analyzed by $^1$H NMR at pH 0.75 at 25 °C as described in Materials and Methods. Relative concentrations of intermediates were determined from the relative integrals of the corresponding signals, which were used to calculate all microscopic rate constants of the catalytic cycle.

**Table 2.** Net rate constants ($s^{-1}$) for reaction of BFDC and PDC variants with benzoylformate and pyruvate.

| Enzyme Variant | Substrate | $k_{cat}$ | C–C ($k_2'$) | Decarboxylation ($k_3'$) | Release ($k_4'$) |
|---|---|---|---|---|---|
| *wt*BFDC [a] | BF | 450 | 500 | 16,000 | 2400 |
| | Pyruvate | <0.05 [b] | ND [c] | ND [c] | ND [c] |
| BFDC T377L/A460Y | BF | 15 | 40 | 500 | 24 |
| | Pyruvate | 3.3 | 6 | 24 | 10 |
| *Sc*PDC [d] | Pyruvate | 38 | 294 | 105 | 105 |
| *Zm*PDC [e] | Pyruvate | 150 | 2650 | 397 | 265 |

[a] Values for *wt*BFDC with BF were obtained from Bruning et al. [25]; [b] Value for *wt*BFDC with pyruvate was obtained from Yep and McLeish [30]; [c] Not determined; [d] Values for *Sc*PDC were obtained from Joseph et al. [33]; [e] Values for *Zm*PDC were obtained from Tittmann et al. [24].

ALThDP arises during an off-pathway reaction in which a second molecule of pyruvate acts as the electron acceptor for the post decarboxylation carbanion intermediate. ALThDP has previously been observed during the determination of the microscopic rate constants for acetohydroxyacid acid synthase [34–36]. The presence of ALThDP during the reaction cycle of the BFDC T377L/A460Y variant suggests that the reaction may generate acetolactate. However, the only carboligation product detected was acetoin. While this seems at odds with the presence of ALThDP, Beigi et al. [37] recently described an α-hydroxy-β-keto acid rearrangement-decarboxylation reaction in which acetolactate spontaneously decarboxylates to yield acetoin. Currently, it is unclear if ALThDP decarboxylates to yield acetoin or if the decarboxylation of acetolactate occurs after it is released. Regardless, it would seem that the longer lifetime of the pyruvate-derived carbanion/enamine results in an increase in formation of a carboligation product.

If this off-pathway reaction is ignored, then the calculation of the forward net rate constants is possible. These are provided in Table 2. Examination of those rate constants shows that C–C bond formation ($k_2'$) is the slowest step in the reaction of BFDC T377L/A460Y with pyruvate. Further, while the variant is clearly more efficient in binding pyruvate, the rate constant for the formation of the LThDP ($6\,\mathrm{s}^{-1}$) is still 50- to 400-fold lower than those of the true PDCs. By comparison, decarboxylation of the LThDP intermediate ($k_3'$) is ca. four-fold faster, and the subsequent steps involving enamine protonation/product release ($k_4'$) are, at best, partially rate-limiting. These results are in sharp contrast with those found in previous studies with *Zm*PDC [24] and *Sc*PDC [33] where formation of LThDP was the fast step and both decarboxylation and product release were appreciably slower (Table 2). That said, it is not without precedent, as initial C–C bond formation ($k_2'$) is the slow step in the reaction of acetohydroxyacid synthase (AHAS) with pyruvate [32]. Finally, it is notable that decarboxylation of LThDP catalyzed by BFDC T377L/A460Y is the fast step in the overall process, and only approximately four-fold slower when compared to *Sc*PDC. It seems that, even with a relatively poor substrate, the active site of BFDC seems to favor decarboxylation over the other reaction steps.

*2.4. X-ray Structure of BFDC T377L/A460Y*

In an attempt to understand how the BFDC T377L/A460Y variant resulted in a better binding site for pyruvate and, potentially, to explain the decrease in the rate of decarboxylation of benzoylformate, the X-ray structure for T377L/A460Y was determined. The structure was solved to a resolution of 1.56 Å and the coordinates have been deposited with the Protein Data Bank as PDB ID 4MZX. The refinement statistics are provided in Table 3.

The overall fold was found to be almost identical to that of *wt*BFDC holoenzyme, and both the space group and cell dimensions were also the same. There was a RMSD deviation of 0.28 Å between the BFDC T377L/A460Y variant and *wt*BFDC enzyme for the Cα atoms of a single monomer. As previously observed with BFDC structures, the region comprising residues 461–470 had the highest B-factors found in the protein.

During the initial rounds of refinement, positive electron density adjacent to the C2 atom of the thiazolium ring was observed when ThDP was modeled into the active site. This was indicative of the oxidation of ThDP to thiamin thiazolone diphosphate (TZD). Consequently TZD was modeled into the active site during subsequent rounds of refinement. This oxidation of ThDP to TZD has been observed previously in BFDC, as well as other ThDP-dependent enzymes, and has been attributed to radiation damage [38,39]. As seen in several earlier BFDC structures, the diphosphate tail of TZD is coordinated to $Ca^{2+}$, rather than $Mg^{2+}$. Presumably due to its high concentrations in the crystallization buffer, $Mg^{2+}$ is often displaced by $Ca^{2+}$ during the crystallization of BFDC [17,38]. In addition to the cofactors, electron density indicative of two glycerol molecules was also observed in the asymmetric unit. One glycerol was H-bonded to the side chains of Arg294 and Asp312, while the second was coordinated to the oxygen atoms of the backbone carbonyl of Asp477 and the side chain of Gln443. Finally, a sodium ion was found coordinated between the main chain carbonyl oxygens of Leu118, Arg120, and Asn117 from two separate monomers at the dimer–dimer interface. The BFDC T377L/A460Y structure retained the backbone conformation of the *wt*BFDC enzyme at the points of the mutations, but, not surprisingly, additional electron density was seen in the difference maps. This density corresponded to leucine and tyrosine residues at positions 377 and 460, respectively.

Comparison of the active sites of the *wt*BFDC and the T377L/A460Y variant showed that the side chain of His281 is displaced by more than 1 Å compared to the *wt*BFDC holoenzyme (Figure 3). The only other notable change was Leu110 being found in two rotamer conformations, one of which had not previously been observed in BFDC or its variants [17,18,25,38,40,41]. The net result was that the active site was reduced in volume by ~20%.

Table 3. Data, model, and crystallographic statistics for BFDC T377L/A460Y structure [a].

| Data Collection Statistics BFDC T377L/A460Y (PDB 4MZX) | |
| --- | --- |
| Beam line | APS, GM/CA-CAT, 23 ID-B |
| Wavelength | 1.03 Å |
| Space group | I222 |
| Cell constants | a = 80.98 Å; b = 95.56 Å; c = 137.2 Å; $\alpha = \beta = \gamma = 90°$ |
| Total reflections | 75885 |
| Unique reflections | 75885 |
| Resolution limit (Å) | 1.56 |
| Completeness (%) | 99.8 (100) |
| Redundancy | 7.3 (7.2) |
| I/σI | 27.5 (7.3) |
| Rmerge (%) | 5.8 (29) |
| **Refinement statistics** | |
| Resolution range (Å) | 1.56–19.71 |
| Rfree test set size | 2000 |
| Rcryst (%) | 11.7 |
| Rfree (%) | 15.0 |
| No. Atoms | |
| Total | 4547 |
| Protein | 4050 |
| Ligand | 41 |
| Water | 456 |
| Ramachandran | |
| Favored | 98 |
| Allowed | 2 |
| Outliers | 0 |
| B-factors | |
| Protein | 15.1 |
| Ligand | 18.4 |
| Solvent | 26.4 |
| R.m.s. deviations | |
| Bond lengths (Å) | 0.017 |
| Bond angles (°) | 1.50 |

[a] Value in parentheses are for the highest-resolution shell.

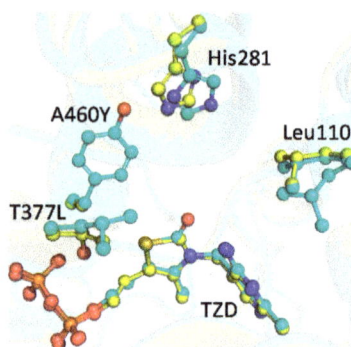

**Figure 3.** Superposition of the active sites of *wt*BFDC (yellow) and the BFDC T377L/A460Y variant (cyan) reveals a shift in the position of His281 accompanied by multiple conformations of Leu110. In addition, ThDP was oxidized to thiamin thiazolone diphosphate (TZD). Figure prepared using PyMOL (Schrödinger, Inc., Portland, OR USA) using data from PDB ID 1BFD (*wt*BFDC) and 4MZX (BFDC T377L/A460Y).

The data in Table 1 show that the $K_m$ value of *wt*BFDC for pyruvate was in excess of 100 mM, and for the T377L/A460Y variant, the $K_m$ value was reduced to <5 mM. Clearly the latter must have a binding site much better able to accommodate pyruvate. To see how this was achieved we compared the binding of the LTHDP to PDC and to BFDC T377L/A460Y. Previously, the *Zm*PDC E473D structure in complex with LThDP was solved [42]. In this structure the methyl substituent of the lactyl moiety is within 5 Å of 11 atoms evenly distributed across seven different residues.

Superimposition of the LThDP intermediate on ThDP in the active site of *wt*BFDC revealed only three atoms within 5 Å of the methyl group (Figure 4A). By contrast, superimposition of LThDP on TZD in the T377L/A460Y structure (Figure 4B) shows that this variant now has 10 atoms contributing to the methyl-binding site. Thus the active site has become more like that of *Zm*PDC. There were some differences for, unlike the evenly distributed interactions which make up the *Zm*PDC methyl-binding pocket, 7 of the 10 potential interactions in the BFDC T377L/A460Y variant arise from the engineered residues, Leu377 and Tyr460. This suggests that changes in other active site residues may be able to further improve pyruvate binding and, potentially, in positioning the "new" substrate for decarboxylation and subsequent steps.

While this relatively simplistic modeling of LThDP offers an explanation for how the active site of the T377L/A460Y variant has improved pyruvate binding, there are still several issues to be addressed. For example, while the active site volume has been reduced by 20% from *wt*BFDC, the $K_m$ value for benzoylformate is only increased by a factor of two. Indeed, at 0.45 mM, it remains 10-fold lower than that of pyruvate (Table 2). To understand how this could be achieved, the tetrahedral intermediate from the reaction of *wt*BFDC with methyl benzoylphosphonate (MBP) [41] was superimposed on the TZD in the active site of T377L/A460Y. The MBP-ThDP intermediate is an analogue of MThDP, thus can be used to get an indication of potential changes in benzoylformate binding.

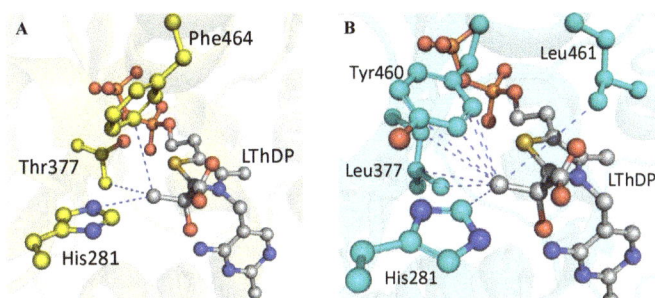

**Figure 4.** (**A**) *wt*BFDC (PDB: 1BFD) with LThDP (gray) placed into the active site by superposition of the LThDP on the corresponding atoms of ThDP. In total, three atoms are within 5 Å of the methyl group of LThDP; (**B**) BFDC T377L/A460Y PDB 4MZX) with LThDP (gray) superimposed on TZD. In total 10 atoms, from four residues, are within 5 Å of the methyl group of LThDP. In both structures blue lines indicate distances of <5 Å. Figure prepared using PyMOL with LThDP data from PDB ID 3OE1 [42].

This simple model indicates a potentially severe steric clash between the phenyl group of the inhibitor and the engineered mutations (Figure 5). The $\gamma$-methyl of Leu377 is within <2.5 Å of several carbon atoms of the phenyl ring of MBP, whereas Tyr460 has a potential edge-to-edge contact with MBP that is also within <2.5 Å. While this appears to be at odds with the kinetic data indicating only a modest effect on $K_m$ value for benzoylformate, there is structural evidence that *wt*BFDC is able to accommodate unnaturally large substrates by slight rotations of active site residues [43]. Therefore it is not unreasonable to suggest that the side chains of Tyr460 and Leu377 could rotate to avoid or, at least, ameliorate unfavorable interactions with the MThDP. Equally plausible is that the phenyl group of the MThDP could rotate to prevent steric clash with these residues. Either scenario provides a rationale for the binding of benzoylformate being relatively unaffected.

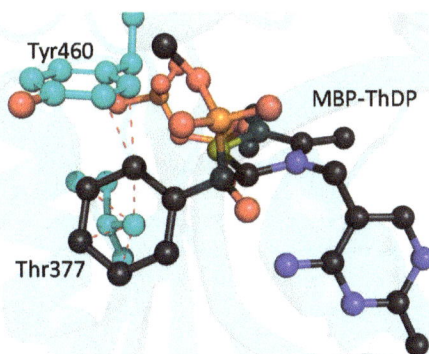

**Figure 5.** BFDC T377L/A460Y with MBP-ThDP (black) placed into the active site. Red lines indicate distances of <2.5 Å. Superposition carried out with PyMOL as described in Materials and Methods using MBP-THDP data from PDB ID 3FSJ [41]

Figure 5 may also provide a partial explanation for the reduction in net rate constants for the decarboxylation of benzoylformate (Table 2). It is clear that some active site rearrangements will be necessary to avoid potential steric clashes and permit substrate binding. Conceivably, these may make it more difficult for the ThDP ylid to attack the carbonyl of benzoylformate, thereby explaining the observed 12-fold decrease in the rate of formation of the MThDP intermediate ($k_2'$).

Figure 3 shows that the mutations have caused modest, but potentially significant changes in the orientation of Leu110 and His281. Together these changes provide some insight into the considerable decrease in the observed rate for the decarboxylation of the MThDP intermediate ($k_3'$), as well as the release of benzaldehyde (Scheme 1). While its precise role in catalysis is not obvious, mutagenesis and crystallographic data suggest that Leu110 plays a vital role in the mechanism of BFDC. The L110A mutation results in greater than four-orders of magnitude decrease in $k_{cat}/K_m$, due to a simultaneous increase in $K_m$ and decrease in $k_{cat}$ values [44]. Analysis of the structure of *wt*BFDC in complex with MBP indicates that Leu110 may be responsible for positioning the glyoxylate moiety of benzoylformate [25,41]. In the T377L/A460Y variant, Leu110 adopts multiple conformations potentially reducing its ability to lock the glyoxylate moiety into the perpendicular arrangement of the carboxylate group to the thiazolium-C2α bond. This geometry is thought to promote decarboxylation by allowing maximum overlap of π electrons of the thiazolium ring and the *p*-orbital of the scissile bond [25,45]. The conformation has been observed in the X-ray structures of all the first tetrahedral intermediates, and intermediate analogues [25,42,46–48]. Further, Ser26 has been proposed to assist in the decarboxylation step by carrying out a nucleophilic attack on carboxylate group of MThDP [40]. In *wt*BFDC, not only does the perpendicular arrangement assure maximum orbital overlay, but also it positions the hydroxyl group of Ser26 within striking distance of the carboxylate group. Taken together, the alternative conformations of Leu110 observed may help explain the decrease in the rate of MThDP decarboxylation ($k_3'$).

The net rate constant associated with product release ($k_4'$) incorporates the protonation of the enamine/carbanion intermediate, as well as the breakdown of the resultant tetrahedral intermediate and subsequent release of benzaldehyde. The T377L/A460Y variant exhibits a 100-fold decrease in $k_4'$, i.e., the largest decrease of all net rate constants. Mutagenesis studies suggest that His281 is involved in the protonation of the enamine/carbanion intermediate [18], whereas His70 has been implicated in the abstraction of the proton from the HBzThDP intermediate, thereby facilitating product release [18,19]. Since these two steps of the mechanism are incorporated into this single net rate constant, it is difficult to say which step has been more affected in the T377L/A460Y variant. The X-ray structure shows that the position of His70 is largely unaffected by the mutations but that there is a 1 Å displacement of the imidazole ring of His281 (Figure 3). It is conceivable that this change could reduce proton transfer efficiency, resulting in an increase in the lifetime of the carbanion/enamine. It is well known that the BFDC H281A variant has been associated with an improvement in carboligation efficiency [49–51], which was attributed to the longer lifetime of the enamine. It is also notable that an increase in the lifetime of the carbanion/enamine in *Zm*PDC gave rise to formation of acetoin/acetolactate [52]. Given that the $^1$H NMR spectrum of the intermediates in the reaction of T377L/A460Y with pyruvate (Figure 2) showed the presence of 2-acetolactate-ThDP, it is reasonable to speculate that displacement of His281 is contributing to the decrease in $k_4'$. That said, there is a clear increase in the steady-state concentration of HBzThDP, which certainly indicates that its breakdown and product release—i.e., $k_5'$ (Scheme 1)—also must be greatly reduced.

## 3. Materials and Methods

### 3.1. Materials

Plasmids containing genes for recombinant alcohol dehydrogenase from *Equus caballus* (HLADH) and C-terminally His$_6$-tagged pyruvate decarboxylase from *Saccharomyces cerevisiae* were kind gifts from Bryce Plapp (University of Iowa) and Frank Jordan (Rutgers University), respectively. NADH, IPTG, yeast alcohol dehydrogenase (YADH), and the various 2-keto acids were purchased from Sigma-Aldrich (St Louis, MO, USA). Other assay reagents were purchased from either Sigma-Aldrich or Fisher Scientific (Waltham, MA, USA) and were of the highest commercially available grade.

### 3.2. Protein Expression and Purification

Expression and purification of *wt*BFDC and the BFDC T377L/A460Y variant was carried out as described in Yep et al. [30]. The protein was exchanged into storage buffer (100 mM KPO$_4$, 1 mM MgSO$_4$, 0.5 mM ThDP, pH 6.0, 10% glycerol *v/v*), and concentrated using EMD Millipore Amicon Ultra filters (Fisher Scientific). Expression and purification of *Sc*PDC was carried out in a similar manner. In all cases purity was assessed by SDS-PAGE, and the protein concentration was determined by the Bradford method [53] using BSA as the standard.

### 3.3. Analysis of Substrate Spectrum

The activity assays with the purified enzymes have been described in detail elsewhere [30]. The assay contained the appropriate 2-keto acid substrate, NADH, and HLADH (YADH for short straight-chain aliphatic substrates) in 100 mM potassium phosphate buffer (pH 6.0) with a final volume of 1 mL. The reaction was carried out at 30 °C and was initiated by the addition of the enzyme. Kinetic data were fitted to the Michaelis-Menten equation using SigmaPlot 9.0.1.

### 3.4. Steady State Analysis of Reaction Intermediates by Acid Quench/NMR Spectroscopy

Prior to the chemical quench/NMR studies, excess ThDP was removed from the BFDC T377L/A460Y sample by dialysis overnight against 50 mM potassium phosphate buffer, pH 6.5, supplemented with 2.5 mM MgSO$_4$. Enzyme concentrations were adjusted to 15 mg/mL and reacted with one equivalent volume of either 50 mM benzoylformate or 100 mM pyruvate prepared in the same buffer as the enzyme. After 1–10 seconds the reaction was quenched by manual mixing with the addition of 1 volume of an acidic solution comprising 65% (*v/v*) D$_2$O, 25% (*v/v*) of trichloroacetic acid and 10% (*v/v*) concentrated HCl [24,25]. Samples were vortexed to facilitate complete quenching and protein denaturation. Samples were then centrifuged and the supernatant passed through a 0.22 μm filter. The filtrate, containing the intermediates, substrates and products, was analyzed by $^1$H NMR spectroscopy. NMR data collection, assignment of peaks, and calculation of net forward unimolecular rate constants was performed as previously described [24,25].

### 3.5. Crystallization of BFDC T377L/A460Y

Crystals of the BFDC T377L/A460Y variant were grown by the hanging drop diffusion method under the conditions used for the crystallization of *wt*BFDC [17]. Storage buffer was exchanged for crystallization buffer (0.1 mM MgCl$_2$, 0.2 mM ThDP, 25 mM NaHEPES pH 7.0). The well solution consisted of 0.1 M Tris (pH 8.5), 0.15 M CaCl$_2$ and 22% PEG 400 (*v/v*). Equal volumes of protein solution (10 mg/mL) and well solution were pipetted onto a silanized glass slide and mixed. A heavy precipitate quickly formed and single crystals emerged from the precipitate within days. Crystals were transferred to fresh crystallization buffer containing 36% glycerol (*v/v*) as a cryoprotectant and mounted on Hampton CryoLoops immediately prior to being flash-frozen in liquid nitrogen.

### 3.6. X-ray Data Collection

Diffraction experiments were carried out at 100 K on the 23-ID-B beamline administered by GM/CA-CAT at the Advanced Photon Source at Argonne National Laboratory. Datasets for the BFDC T377L/A460Y variant were scaled to the $I_{222}$ space group. Data reduction and processing of dataset was done with HKL2000 software package and the CCP4 suite of programs [54]. Molecular replacement was performed using the search model *wt*BFDC (PDB 1BFD) with metals and waters removed. The asymmetric unit for each variant contained a single monomer.

### 3.7. Structure Solutions and Refinements

Refinement was performed using Phenix.refine [55]. After each round of refinement, the electron density was manually inspected and models were built using Coot [56]. Model refinement continued until the free $R$ factor and the crystallographic $R$ factor had converged. Validity of the model was checked using the Molprobity server [57,58]. All images were generated with PyMOL (Schrödinger Inc.).

*3.8. Modeling of Reaction Analogues and Intermediates*

Using PyMOL and the coordinates from PDB ID 3OE1, the lactyl-ThDP intermediate was placed into the active sites of both *wt*BFDC (PDB ID: 1BFD) and the T377L/A460Y structure. Further, using the coordinates from PDB ID 3FZN, the phosphonomandelyl-ThDP (MBP-ThDP) was also placed into the active site of the T377L/A460Y structure. In this process, the Cα atoms of the protein structures were first superimposed. Next, the atoms corresponding to the ThDP moiety of lactyl-ThDP and MBP-ThDP were aligned with the corresponding atoms of either ThDP, found in *wt*BFDC, or the thiamin thiazolone diphosphate (TZD) found in the BFDC T377L/A460Y structure. No energy minimization was carried out.

## 4. Conclusions

The BFDC variant T377L/A460Y effectively acts as a "pseudo" PDC. This is reflected in a broad substrate spectrum more akin to that of *Sc*PDC than *wt*BFDC. NMR-based analysis of the steady-state reaction intermediates for the reaction of BFDC T377L/A460Y with pyruvate revealed that the rate-limiting step for this variant is the formation of the LThDP intermediate. This is in contrast to the reactions catalyzed by *Zm*PDC and *Sc*PDC, where product release is rate-limiting. Comparing the methyl-binding pocket of T377L/A460Y to *Zm*PDC suggests that both are likely to have a similar number of interactions with the lactyl moiety. However, the contacts within the *Zm*PDC methyl-binding pocket are evenly distributed across seven residues compared to only four residues in T377L/A460Y. Further, the seven residues of *Zm*PDC are arranged to surround the methyl group thereby providing steric restraints from every direction. Presumably this arrangement would also allow for the maximum orbital overlap, thereby permitting attack of the ThDP ylid and facilititating the subsequent decarboxylation of the LThDP intermediate. On the other hand, the four residues comprising the T377L/A460Y methyl binding pocket are unevenly arranged and do not provide the precise substrate positioning seen in a native PDC. This likely accounts for the higher $K_m$ value for pyruvate, and the lower rate of formation of LThDP observed with the BFDC variant. Given that several aliphatic 2-keto acids are better substrates than pyruvate for T377L/A460Y, it is reasonable to predict that further evolution could well bring the properties of the "pseudo" PDC significantly closer to those of the native enzyme.

Over the years there have been many attempts to interconvert or otherwise alter substrate specificity in ThDP-dependent enzymes. This has been met with a resounding lack of success [1]. The results here provide some rationale for the difficulty. They highlight the fact that, in spite of the similarity in overall mechanism, ThDP-dependent enzymes often have different rate-determining steps. This means that the enzymes have evolved so that individual residues will provide an optimal alignment for (effectively) several substrates (intermediates) in the catalytic cycle. Thus, making a change that may be expected to improve binding for a new substrate may have a deleterious effect on another step in the mechanism. In the BFDC T377L/A460Y variant, for example, not only do we see a new ability to catalyze pyruvate decarboxylation, but it is done without greatly affecting its ability to bind benzoylformate. Rather, the decrease in $k_{cat}$ value for reaction with benzoylformate is due primarily to a change in rate-determining step. This result implies that structure-based redesign of ThDP-dependent enzymes will prove difficult. Rather, a more random approach will be required, probably involving saturation mutagenesis at a combination of sites identified on the basis of X-ray structures or homology models.

**Acknowledgments:** This work was supported by the National Science Foundation (CHE 1306877to M.J.M.) and the Deutsche Forschungsgemeinschaft (FOR1296 to K.T.). Use of the Advanced Photon Source, an Office of Science User Facility operated for the U.S. Department of Energy (DOE) Office of Science by Argonne National

Laboratory, was supported by the U.S. DOE under Contract DE-AC02-06CH11357. GM/CA @ APS has been funded in whole or in part with Federal funds from the National Cancer Institute (Y1-CO-1020) and the National Institute of General Medical Sciences (Y1-GM-1104).

**Author Contributions:** F.H.A. and M.J.M conceived and designed the experiments; F.H.A and M.P.R carried out the enzyme kinetic experiments; F.H.A., C.W. and D.M. carried out the NMR experiments while K.T. analyzed the NMR data; F.H.A carried out the crystallography experiments and analyzed the data; F.H.A and M.J.M. wrote the paper.

**Conflicts of Interest:** The authors declare no conflict of interest.

## References

1. Andrews, F.H.; McLeish, M.J. Substrate specificity in thiamin diphosphate-dependent decarboxylases. *Bioorg. Chem.* **2012**, *43*, 26–36. [CrossRef] [PubMed]
2. Nichols, N.; Dien, B.; Bothast, R. Engineering lactic acid bacteria with pyruvate decarboxylase and alcohol dehydrogenase genes for ethanol production from *Zymomonas mobilis*. *J. Ind. Microbiol. Biotechnol.* **2003**, *30*, 315–321. [CrossRef] [PubMed]
3. Dyda, F.; Furey, W.; Swaminathan, S.; Sax, M.; Farrenkopf, B.; Jordan, F. Catalytic centers in the thiamin diphosphate dependent enzyme pyruvate decarboxylase at 2.4 Å resolution. *Biochemistry* **1993**, *32*, 6165–6170. [CrossRef] [PubMed]
4. Dobritzsch, D.; König, S.; Schneider, G.; Lu, G. High resolution crystal structure of pyruvate decarboxylase from *Zymomonas mobilis*. Implications for substrate activation in pyruvate decarboxylases. *J. Biol. Chem.* **1998**, *273*, 20196–20204. [CrossRef] [PubMed]
5. Kutter, S.; Wille, G.; Relle, S.; Weiss, M.S.; Hübner, G.; König, S. The crystal structure of pyruvate decarboxylase from *Kluyveromyces lactis*. *FEBS J.* **2006**, *273*, 4199–4209. [CrossRef] [PubMed]
6. Rother, D.; Kolter, G.; Gerhards, T.; Berthold, C.L.; Gauchenova, E.; Knoll, M.; Pleiss, J.; Müller, M.; Schneider, G.; Pohl, M. S-selective mixed carboligation by structure-based design of the pyruvate decarboxylase from *Acetobacter pasteurianus*. *ChemCatChem* **2011**, *3*, 1587–1596. [CrossRef]
7. Liu, M.; Sergienko, E.A.; Guo, F.; Wang, J.; Tittmann, K.; Hübner, G.; Furey, W.; Jordan, F. Catalytic acid-base groups in yeast pyruvate decarboxylase. 1. Site-directed mutagenesis and steady-state kinetic studies on the enzyme with the D28A, H114F, H115F, and E477Q substitutions. *Biochemistry* **2001**, *40*, 7355–7368. [CrossRef] [PubMed]
8. Sergienko, E.A.; Jordan, F. Catalytic acid-base groups in yeast pyruvate decarboxylase. 2. Insights into the specific roles of D28 and E477 from the rates and stereospecificity of formation of carboligase side products. *Biochemistry* **2001**, *40*, 7369–7381. [CrossRef] [PubMed]
9. Sergienko, E.A.; Jordan, F. Catalytic acid-base groups in yeast pyruvate decarboxylase. 3. A steady-state kinetic model consistent with the behavior of both wild-type and variant enzymes at all relevant pH values. *Biochemistry* **2001**, *40*, 7382–7403. [CrossRef] [PubMed]
10. Candy, J.M.; Duggleby, R.G. Structure and properties of pyruvate decarboxylase and site-directed mutagenesis of the *Zymomonas mobilis* enzyme. *BBA Protein Struct. Mol. Enzymol.* **1998**, *1385*, 323–338. [CrossRef]
11. Schütz, A.; Sandalova, T.; Ricagno, S.; Hübner, G.; König, S.; Schneider, G. Crystal structure of thiamindiphosphate-dependent indolepyruvate decarboxylase from *Enterobacter cloacae*, an enzyme involved in the biosynthesis of the plant hormone indole-3-acetic acid. *Eur. J. Biochem.* **2003**, *270*, 2312–2321. [CrossRef] [PubMed]
12. Berthold, C.L.; Gocke, D.; Wood, M.D.; Leeper, F.J.; Pohl, M.; Schneider, G. Structure of the branched-chain keto acid decarboxylase (KdcA) from *Lactococcus lactis* provides insights into the structural basis for the chemoselective and enantioselective carboligation reaction. *Acta Crystallogr. D Biol. Crystallogr.* **2007**, *63*, 1217–1224. [CrossRef] [PubMed]
13. Versees, W.; Spaepen, S.; Vanderleyden, J.; Steyaert, J. The crystal structure of phenylpyruvate decarboxylase from *Azospirillum brasilense* at 1.5 Å resolution. Implications for its catalytic and regulatory mechanism. *FEBS J.* **2007**, *274*, 2363–2375. [CrossRef] [PubMed]
14. Hegeman, G.D. Benzoylformate decarboxylase (*Pseudomonas putida*). *Methods Enzymol.* **1970**, *17*, 674–678.
15. Hegeman, G.D. Synthesis of the enzymes of the mandelate pathway by *Pseudomonas putida* I. Synthesis of enzymes by the wild type. *J. Bacteriol.* **1966**, *91*, 1140–1154. [PubMed]

16. Kluger, R.; Tittmann, K. Thiamin diphosphate catalysis: Enzymic and nonenzymic covalent intermediates. *Chem. Rev.* **2008**, *108*, 1797–1833. [CrossRef] [PubMed]
17. Hasson, M.S.; Muscate, A.; McLeish, M.J.; Polovnikova, L.S.; Gerlt, J.A.; Kenyon, G.L.; Petsko, G.A.; Ringe, D. The crystal structure of benzoylformate decarboxylase at 1.6 Å resolution: Diversity of catalytic residues in thiamin diphosphate-dependent enzymes. *Biochemistry* **1998**, *37*, 9918–9930. [CrossRef] [PubMed]
18. Polovnikova, E.S.; McLeish, M.J.; Sergienko, E.A.; Burgner, J.T.; Anderson, N.L.; Bera, A.K.; Jordan, F.; Kenyon, G.L.; Hasson, M.S. Structural and kinetic analysis of catalysis by a thiamin diphosphate-dependent enzyme, benzoylformate decarboxylase. *Biochemistry* **2003**, *42*, 1820–1830. [CrossRef] [PubMed]
19. Sergienko, E.A.; Wang, J.; Polovnikova, L.; Hasson, M.S.; McLeish, M.J.; Kenyon, G.L.; Jordan, F. Spectroscopic detection of transient thiamin diphosphate-bound intermediates on benzoylformate decarboxylase. *Biochemistry* **2000**, *39*, 13862–13869. [CrossRef] [PubMed]
20. Schütz, A.; Golbik, R.; König, S.; Hübner, G.; Tittmann, K. Intermediates and transition states in thiamin diphosphate-dependent decarboxylases. A kinetic and NMR study on wild-type indolepyruvate decarboxylase and variants using indolepyruvate, benzoylformate, and pyruvate as substrates. *Biochemistry* **2005**, *44*, 6164–6179. [CrossRef] [PubMed]
21. Kern, D.; Kern, G.; Neef, H.; Tittmann, K.; Killenberg-Jabs, M.; Wikner, C.; Schneider, G.; Hübner, G. How thiamin diphosphate is activated in enzymes. *Science* **1997**, *275*, 67–70. [CrossRef] [PubMed]
22. Frank, R.; Leeper, F.; Luisi, B. Structure, mechanism and catalytic duality of thiamine-dependent enzymes. *Cell Mol. Life Sci.* **2007**, *64*, 892–905. [CrossRef] [PubMed]
23. Meyer, D.; Neumann, P.; Ficner, R.; Tittmann, K. Observation of a stable carbene at the active site of a thiamin enzyme. *Nat. Chem. Biol.* **2013**, *9*, 488–490. [CrossRef] [PubMed]
24. Tittmann, K.; Golbik, R.; Uhlemann, K.; Khailova, L.; Schneider, G.; Patel, M.; Jordan, F.; Chipman, D.M.; Duggleby, R.G.; Hübner, G. NMR analysis of covalent intermediates in thiamin diphosphate enzymes. *Biochemistry* **2003**, *42*, 7885–7891. [CrossRef] [PubMed]
25. Bruning, M.; Berheide, M.; Meyer, D.; Golbik, R.; Bartunik, H.; Liese, A.; Tittmann, K. Structural and kinetic studies on native intermediates and an intermediate analogue in benzoylformate decarboxylase reveal a least motion mechanism with an unprecedented short-lived predecarboxylation intermediate. *Biochemistry* **2009**, *48*, 3258–3268. [CrossRef] [PubMed]
26. Pohl, M.; Lingen, B.; Müller, M. Thiamin-diphosphate-dependent enzymes: New aspects of asymmetric C–C bond formation. *Chem. Eur. J.* **2002**, *8*, 5288–5295. [CrossRef]
27. Fuganti, C.; Grasselli, P. Synthesis of the C$_{14}$ chromanyl moiety of natural α-tocopherol (vitamin E). *J. Chem. Soc. Chem. Commun.* **1982**, *4*, 205–206. [CrossRef]
28. Gala, D.; DiBenedetto, D.J.; Clark, J.E.; Murphy, B.L.; Schumacher, D.P.; Steinman, M. Preparations of antifungal Sch 42427/SM 9164: Preparative chromatographic resolution, and total asymmetric synthesis via enzymic preparation of chiral α-hydroxy arylketones. *Tetrahedron Lett.* **1996**, *37*, 611–614. [CrossRef]
29. Siegert, P.; McLeish, M.J.; Baumann, M.; Iding, H.; Kneen, M.M.; Kenyon, G.L.; Pohl, M. Exchanging the substrate specificities of pyruvate decarboxylase from *Zymomonas mobilis* and benzoylformate decarboxylase from *Pseudomonas putida*. *Protein Eng. Des. Sel.* **2005**, *18*, 345–357. [CrossRef] [PubMed]
30. Yep, A.; McLeish, M.J. Engineering the substrate binding site of benzoylformate decarboxylase. *Biochemistry* **2009**, *48*, 8387–8395. [CrossRef] [PubMed]
31. Tittmann, K.; Vyazmensky, M.; Hübner, G.; Barak, Z.; Chipman, D.M. The carboligation reaction of acetohydroxyacid synthase ii: Steady-state intermediate distributions in wild type and mutants by nmr. *Proc Natl Acad Sci U S A* **2005**, *102*, 553–558. [CrossRef] [PubMed]
32. Belenky, I.; Steinmetz, A.; Vyazmensky, M.; Barak, Z.; Tittmann, K.; Chipman, D.M. Many of the functional differences between acetohydroxyacid synthase (AHAS) isozyme I and other AHASs are a result of the rapid formation and breakdown of the covalent acetolactate-thiamin diphosphate adduct in AHAS I. *FEBS J.* **2012**, *279*, 1967–1979. [CrossRef] [PubMed]
33. Joseph, E.; Wei, W.; Tittmann, K.; Jordan, F. Function of a conserved loop of the β-domain, not involved in thiamin diphosphate binding, in catalysis and substrate activation in yeast pyruvate decarboxylase. *Biochemistry* **2006**, *45*, 13517–13527. [CrossRef] [PubMed]
34. Vinogradov, M.; Kaplun, A.; Vyazmensky, M.; Engel, S.; Golbik, R.; Tittmann, K.; Uhlemann, K.; Meshalkina, L.; Barak, Z.; Hübner, G.; et al. Monitoring the acetohydroxy acid synthase reaction and

related carboligations by circular dichroism spectroscopy. *Anal. Biochem.* **2005**, *342*, 126–133. [CrossRef] [PubMed]

35. Steinmetz, A.; Vyazmensky, M.; Meyer, D.; Barak, Z.; Golbik, R.; Chipman, D.M.; Tittmann, K. Valine 375 and phenylalanine 109 confer affinity and specificity for pyruvate as donor substrate in acetohydroxy acid synthase isozyme II from *Escherichia coli. Biochemistry* **2010**, *49*, 5188–5199. [CrossRef] [PubMed]

36. Vyazmensky, M.; Steinmetz, A.; Meyer, D.; Golbik, R.; Barak, Z.; Tittmann, K.; Chipman, D.M. Significant catalytic roles for Glu47 and Gln110 in all four of the C–C bond-making and -breaking steps of the reactions of acetohydroxyacid synthase II. *Biochemistry* **2011**, *50*, 3250–3260. [CrossRef] [PubMed]

37. Beigi, M.; Loschonsky, S.; Lehwald, P.; Brecht, V.; Andrade, S.L.A.; Leeper, F.J.; Hummel, W.; Muller, M. α-hydroxy-β-keto acid rearrangement-decarboxylation: Impact on thiamine diphosphate-dependent enzymatic transformations. *Org. Biomol. Chem.* **2013**, *11*, 252–256. [CrossRef] [PubMed]

38. Andrews, F.H.; Tom, A.R.; Gunderman, P.R.; Novak, W.R.P.; McLeish, M.J. A bulky hydrophobic residue is not required to maintain the V-conformation of enzyme-bound thiamin diphosphate. *Biochemistry* **2013**, *52*, 3028–3030. [CrossRef] [PubMed]

39. Berthold, C.L.; Moussatche, P.; Richards, N.G.; Lindqvist, Y. Structural basis for activation of the thiamin diphosphate-dependent enzyme oxalyl-coA decarboxylase by adenosine diphosphate. *J. Biol. Chem.* **2005**, *280*, 41645–41654. [CrossRef] [PubMed]

40. Bera, A.K.; Polovnikova, L.S.; Roestamadji, J.; Widlanski, T.S.; Kenyon, G.L.; McLeish, M.J.; Hasson, M.S. Mechanism-based inactivation of benzoylformate decarboxylase, a thiamin diphosphate-dependent enzyme. *J. Am. Chem. Soc.* **2007**, *129*, 4120–4121. [CrossRef] [PubMed]

41. Brandt, G.S.; Kneen, M.M.; Chakraborty, S.; Baykal, A.T.; Nemeria, N.; Yep, A.; Ruby, D.I.; Petsko, G.A.; Kenyon, G.L.; McLeish, M.J.; et al. Snapshot of a reaction intermediate: Analysis of benzoylformate decarboxylase in complex with a benzoylphosphonate inhibitor. *Biochemistry* **2009**, *48*, 3247–3257. [CrossRef] [PubMed]

42. Meyer, D.; Neumann, P.; Parthier, C.; Friedemann, R.; Nemeria, N.; Jordan, F.; Tittmann, K. Double duty for a conserved glutamate in pyruvate decarboxylase: Evidence of the participation in stereoelectronically controlled decarboxylation and in protonation of the nascent carbanion/enamine intermediate. *Biochemistry* **2010**, *49*, 8197–8212. [CrossRef] [PubMed]

43. Chakraborty, S.; Nemeria, N.S.; Balakrishnan, A.; Brandt, G.S.; Kneen, M.M.; Yep, A.; McLeish, M.J.; Kenyon, G.L.; Petsko, G.A.; Ringe, D.; et al. Detection and time course of formation of major thiamin diphosphate-bound covalent intermediates derived from a chromophoric substrate analogue on benzoylformate decarboxylase. *Biochemistry* **2009**, *48*, 981–994. [CrossRef] [PubMed]

44. Andrews, F.H.; Rogers, M.P.; Paul, L.N.; McLeish, M.J. Perturbation of the monomer-monomer interfaces of the benzoylformate decarboxylase tetramer. *Biochemistry* **2014**, *53*, 4358–4367. [CrossRef] [PubMed]

45. Turano, A.; Furey, W.; Pletcher, J.; Sax, M.; Pike, D.; Kluger, R. Synthesis and crystal structure of an analog of 2-(alpha-lactyl)thiamin, racemic methyl 2-hydroxy-2-(2-thiamin)ethylphosphonate chloride trihydrate. A conformation for a least-motion, maximum-overlap mechanism for thiamin catalysis. *J. Am. Chem. Soc.* **1982**, *104*, 3089–3095. [CrossRef]

46. Wille, G.; Meyer, D.; Steinmetz, A.; Hinze, E.; Golbik, R.; Tittmann, K. The catalytic cycle of a thiamin diphosphate enzyme examined by cryocrystallography. *Nat. Chem. Biol.* **2006**, *2*, 324–328. [CrossRef] [PubMed]

47. Kale, S.; Arjunan, P.; Furey, W.; Jordan, F. A dynamic loop at the active center of the *Escherichia coli* pyruvate dehydrogenase complex e1 component modulates substrate utilization and chemical communication with the E2 component. *J. Biol. Chem.* **2007**, *282*, 28106–28116. [CrossRef] [PubMed]

48. Arjunan, P.; Sax, M.; Brunskill, A.; Chandrasekhar, K.; Nemeria, N.; Zhang, S.; Jordan, F.; Furey, W. A thiamin-bound, pre-decarboxylation reaction intermediate analogue in the pyruvate dehydrogenase E1 subunit induces large scale disorder-to-order transformations in the enzyme and reveals novel structural features in the covalently bound adduct. *J. Biol. Chem.* **2006**, *281*, 15296–15303. [CrossRef] [PubMed]

49. Demir, A.S.; Dünnwald, T.; Iding, H.; Pohl, M.; Müller, M. Asymmetric benzoin reaction catalyzed by benzoylformate decarboxylase. *Tetrahedron Asymmetry* **1999**, *10*, 4769–4774. [CrossRef]

50. Dünkelmann, P.; Kolter-Jung, D.; Nitsche, A.; Demir, A.S.; Siegert, P.; Lingen, B.; Baumann, M.; Pohl, M.; Müller, M. Development of a donor-acceptor concept for enzymatic cross-coupling reactions of aldehydes: The first asymmetric cross-benzoin condensation. *J. Am. Chem. Soc.* **2002**, *124*, 12084–12085. [CrossRef] [PubMed]

51. Pohl, M.; Sprenger, G.A.; Müller, M. A new perspective on thiamine catalysis. *Curr. Opin. Biotechnol.* **2004**, *15*, 335–342. [CrossRef] [PubMed]
52. Meyer, D.; Walter, L.; Kolter, G.; Pohl, M.; Müller, M.; Tittmann, K. Conversion of pyruvate decarboxylase into an enantioselective carboligase with biosynthetic potential. *J. Am. Chem. Soc.* **2011**, *133*, 3609–3616. [CrossRef] [PubMed]
53. Bradford, M.M. A rapid and sensitive method for the quantitation of microgram quantities of protein utilizing the principle of protein-dye binding. *Anal. Biochem.* **1976**, *72*, 248–254. [CrossRef]
54. Bailey, S. The CCP4 suite—Programs for protein crystallography. *Acta Crystallogr. D Biol. Crystallogr.* **1994**, *50*, 760–763.
55. Adams, P.D.; Afonine, P.V.; Bunkoczi, G.; Chen, V.B.; Davis, I.W.; Echols, N.; Headd, J.J.; Hung, L.W.; Kapral, G.J.; Grosse-Kunstleve, R.W.; et al. Phenix: A comprehensive python-based system for macromolecular structure solution. *Acta Crystallogr. D Biol. Crystallogr.* **2010**, *66*, 213–221. [CrossRef] [PubMed]
56. Emsley, P.; Cowtan, K. Coot: Model-building tools for molecular graphics. *Acta Crystallogr. D Biol. Crystallogr.* **2004**, *60*, 2126–2132. [CrossRef] [PubMed]
57. Davis, I.W.; Leaver-Fay, A.; Chen, V.B.; Block, J.N.; Kapral, G.J.; Wang, X.; Murray, L.W.; Arendall, W.B., 3rd; Snoeyink, J.; Richardson, J.S.; et al. Molprobity: All-atom contacts and structure validation for proteins and nucleic acids. *Nucleic Acids Res.* **2007**, *35*, W375–W383. [CrossRef] [PubMed]
58. Chen, V.B.; Arendall, W.B., 3rd; Headd, J.J.; Keedy, D.A.; Immormino, R.M.; Kapral, G.J.; Murray, L.W.; Richardson, J.S.; Richardson, D.C. Molprobity: All-atom structure validation for macromolecular crystallography. *Acta Crystallogr. D Biol. Crystallogr.* **2010**, *66*, 12–21. [CrossRef] [PubMed]

![catalysts logo] *catalysts*

MDPI

*Article*

# Efficient Production of Enantiopure D-Lysine from L-Lysine by a Two-Enzyme Cascade System

Xin Wang, Li Yang, Weijia Cao, Hanxiao Ying, Kequan Chen * and Pingkai Ouyang

State Key Laboratory of Materials-Oriented Chemical Engineering,
College of Biotechnology and Pharmaceutical Engineering, Nanjing Tech University,
Nanjing 211816, Jiangsu, China; xinwang1988@njtech.edu.cn (X.W.); yangliwang@njtech.edu.cn (L.Y.);
caoweijia1989@njtech.edu.cn (W.C.); hxying@njtech.edu.cn (H.Y.); ouyangpk@njtech.edu.cn (P.O.)
* Correspondence: kqchen@njtech.edu.cn; Tel.: +86-25-5813-9386

Academic Editors: Jose M. Palomo and Cesar Mateo
Received: 24 September 2016; Accepted: 25 October 2016; Published: 30 October 2016

**Abstract:** The microbial production of D-lysine has been of great interest as a medicinal raw material. Here, a two-step process for D-lysine production from L-lysine by the successive microbial racemization and asymmetric degradation with lysine racemase and decarboxylase was developed. The whole-cell activities of engineered *Escherichia coli* expressing racemases from the strains *Proteus mirabilis* (LYR) and *Lactobacillus paracasei* (AAR) were first investigated comparatively. When the strain BL21-LYR with higher racemization activity was employed, L-lysine was rapidly racemized to give DL-lysine, and the D-lysine yield was approximately 48% after 0.5 h. Next, L-lysine was selectively catabolized to generate cadaverine by lysine decarboxylase. The comparative analysis of the decarboxylation activities of resting whole cells, permeabilized cells, and crude enzyme revealed that the crude enzyme was the best biocatalyst for enantiopure D-lysine production. The reaction temperature, pH, metal ion additive, and pyridoxal 5′-phosphate content of this two-step production process were subsequently optimized. Under optimal conditions, 750.7 mmol/L D-lysine was finally obtained from 1710 mmol/L L-lysine after 1 h of racemization reaction and 0.5 h of decarboxylation reaction. D-lysine yield could reach 48.8% with enantiomeric excess (ee) $\geq$ 99%.

**Keywords:** D-lysine; racemase; decarboxylase; two-enzyme cascade system

## 1. Introduction

D-Amino acids are being used increasingly used as a starting raw material in the production of valuable pharmaceuticals [1]. Among them, D-lysine has been employed for the synthesis of luteinizing-hormone-releasing hormone analog, or as a drug carrier in the form of polylysine [2,3]. Nowadays, several chemical or biochemical synthesis methods for D-lysine synthesis have been described [3]. In a previous study, D-lysine was successfully prepared from L-lysine by chemical racemization and microbial asymmetric degradation [4]. However, the chemical racemization of amino acids to prepare DL-amino acids is a complex process and requires severe reaction conditions, such as high temperature, strong acid, or alkali [3,4]. Meanwhile, the chemical resolution of DL-amino acids is inefficient with low optical purity of products, and requires expensive chiral resolving agent. Therefore, an efficient bio-based process for the commercial production of D-lysine is highly desirable.

The enzymatic synthesis of D-amino acids can be performed by hydrolases, or D-amino acid aminotransferases with synthetic intermediates and prochiral substrates as starting materials [5]. Currently, industry has an overcapacity for L-lysine with an annual production of more than 2 million tons [6]. We therefore focused on the use of L-lysine as a raw material for D-lysine production. Considering the current preparation process, a two-step reaction containing racemization and asymmetric degradation of L-lysine was regarded as a simple and economical process to synthesize D-lysine effectively (Figure 1). In the first step, amino acid racemase was employed to catalyze racemization of L-lysine to DL-lysine. Amino acid racemases, which are widely prevalent in many living organisms, are divided into two groups: pyridoxal 5′-phosphate (PLP)-dependent and PLP-independent enzymes [7]. However, lysine racemase, one of the PLP-dependent enzymes, has only been described in a small number of bacteria, such as *Proteus mirabilis* [8] and *Oenococcus oeni* [9]. The heterologous expression of lysine racemase for large-scale production of DL-lysine has rarely been reported.

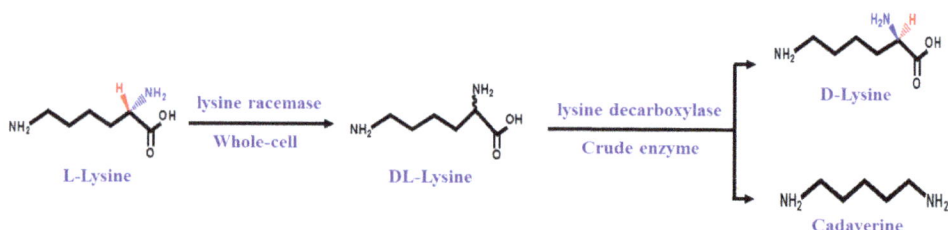

**Figure 1.** Scheme of enantiopure D-lysine production catalyzed by a two-enzyme cascade system of lysine racemase and lysine decarboxylase. The two enzymes were expressed in *Escherichia coli* (*E. coli*) respectively.

In order to obtain enantiopure D-lysine, chiral-selective degradation of L-lysine from the reaction mixture of DL-lysine is necessary. In previous reports, when DL-amino acids were employed as starting materials, D-amino acids, such as D-glutamate, D-arginine, or D-homoserine, have been synthesized by *N*-acyl-D-amino acid amidohydrolase, L-amino acid oxidase, or D-succinylase [10–12]. However, the related enzymes that are active with lysine were only found in few microorganisms, and their catalytic efficiency was too low to meet the demand of the high optical purity of D-lysine [13,14]. Therefore, it was vital to develop an efficient process that degraded L-lysine stereoselectively in the DL-lysine mixture to prepare the enantiopure D-lysine.

In our study, the two racemases from *Proteus mirabilis* (*P. mirabilis*) and *Lactobacillus paracasei* (*L. paracasei*), which have been reported to be active with L-lysine, were heterologously expressed in *Escherichia coli* (*E. coli*), and their activities to generate DL-lysine from L-lysine were compared. Subsequently, L-lysine was selectively catabolized to cadaverine by lysine decarboxylase, and the activities of resting whole cells, permeabilized cells, and crude enzyme were compared to identify the best biocatalyst for enantiopure D-lysine production. After defining the optimal conditions, an efficient bioconversion system for D-lysine production from L-lysine was described by racemization and asymmetric degradation with lysine racemase and decarboxylase, as shown in Figure 1.

## 2. Results

### 2.1. Construction of a Highly Efficient E. coli Whole-Cell Biocatalyst for DL-Lysine Production

In this study, amino racemases from *P. mirabilis* BCRC10725 (LYR) and *L. paracasei* (AAR), which have been reported capable of racemizing L-lysine to D-lysine, were selected [8,15]. The recombinant *E. coli* BL21 (DE3) harboring plasmid pET-28a-LYR or pET-28a-AAR was induced with IPTG (Figure 2a). The recombinant racemase expression was confirmed by SDS-PAGE analysis. It was found that protein LYR (45 kDa) and protein AAR (43 kDa) were all expressed in *E. coli* (Figure 2b), whose molecular weight was consistent with the prediction by gene sequencing. However, protein AAR was found to be highly insoluble in *E. coli*. Next, the specific activities of the two recombinant strains were compared. As shown in Figure 2c, the whole-cell BL21-LYR exhibited higher catalytic conversion activity of L-lysine to product D-lysine than BL21-AAR.

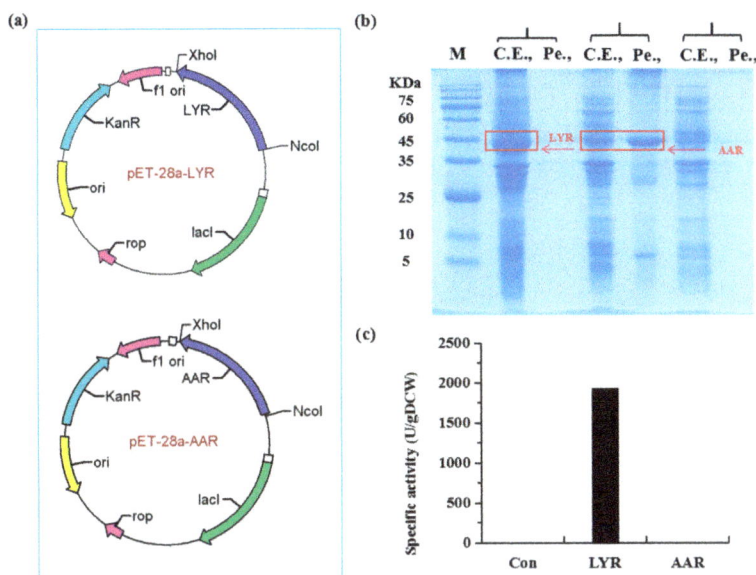

**Figure 2.** Comparison of specific activities of the whole cell BL21-LYR and BL21-AAR. (**a**) The racemases from *Proteus mirabilis* BCRC10725 (lysine racemase, LYR) and *Lactobacillus casei* (amino acid racemase, AAR) were cloned into plasmid pET-28a to generate plasmid pET-28a-LYR and pET-28a-AAR and transformed into *E. coli* BL21 (DE3) respectively; (**b**) Sodium dodecyl sulfate-polyacrylamide gel electrophoresis (SDS-PAGE) of cell extracts of *E. coli* BL21-LYR and BL21-AAR grown in Luria–Bertani (LB) medium at 30 °C. Equal amounts (20 μg) of protein were applied to each lane. Vertical lanes are labeled by M (protein size marker), C.E. (crude extract), or Pe. (pellet); (**c**) The specific activities of the whole cell BL21-LYR and BL21-AAR.

## 2.2. A Two-Step Process for Enantiopure D-Lysine Preparation

Using the whole-cell biocatalyst BL21-LYR, L-lysine was transformed to DL-lysine. Subsequently, to prepare the enantiopure D-lysine, it was necessary to selectively degrade L-lysine to a compound that could be easily separated from D-lysine. In our research, the recombinant *E. coli* ATS3, which has been engineered for the efficient bioconversion of lysine to cadaverine [16], was employed for the asymmetric degradation of L-lysine to cadaverine. As shown in Figure 3, L-lysine was rapidly racemized to generate DL-lysine with a D-lysine yield of 48% at 0.5 h. Next, whole cells of *E. coli* ATS3 (5.0 of $OD_{600}$) were added into the reaction mixture. Within 0.5 h, 85.3% of L-lysine (215 mmol/L) could be converted to cadaverine. However, 14.7% of L-lysine still remained in the reaction solution after 3 h. The enantiomeric excess of D-lysine was only 83%.

To improve the purity, we developed two other processes that permeabilized cells, and crude extracts of the strain ATS3 were used as the biocatalysts to compare their effects on D-lysine enantiomeric excess (Figure 3). Whole cells of AST3 (5.0 of $OD_{600}$) were treated with 0.5% Triton X-100 to obtain permeabilized cells, while the crude enzyme was prepared by sonic disruption. The permeabilized cells were found to exhibit higher activities than whole cells, and the enantiomeric excess of D-lysine reached 92.2% (Figure 3). Then, 0.2 g/L crude enzyme, which was equal to the protein amount in 5.0 of $OD_{600}$ of whole cells, was used to perform the asymmetric degradation experiment. As shown in Figure 3, L-lysine could be completely degraded after 0.5 h, and the enantiomeric excess of D-lysine reached 99.9% with a yield of 46.5%. The decarboxylation reactions were all performed with the addition of 1 mM PLP. Our results demonstrated that the crude enzyme was the best biocatalyst for enantiopure D-lysine production.

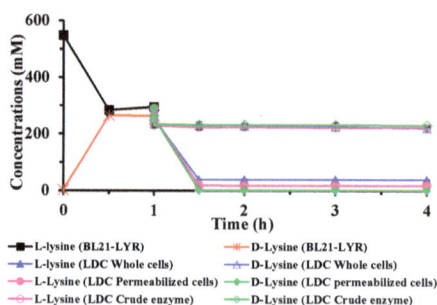

**Figure 3.** Comparison of the decarboxylation activities of resting whole cells, permeabilized cells, and cell extracts of *E. coli* AST3 expressing lysine decarboxylase. The whole-cell BL21-LYR was first employed to racemizing L-lysine to generate DL-lysine. Subsequently, L-lysine in the DL-lysine mixture was degraded by the biocatalysts of *E. coli* ATS3.

## 2.3. Optimization of Reaction Conditions for the Two-Step D-Lysine Production Process

### 2.3.1. Characterization of the Whole-Cell BL21-LYR

To improve the catalytic efficiency of the whole-cell BL21-LYR, it was essential to characterize the effect of the biocatalytic conditions on LYR activity. When the reaction pH ranged from 4.0 to 7.0, it was found that the LYR activity clearly increased with a rise in reaction pH, and reached a maximum at pH 7.0 (Figure 4a). At pH values over 7.0, the LYR activity decreased considerably. To determine the optimal reaction temperature, the bioconversion was carried out at 20, 25, 30, 37, 40, or 45 °C for 10 min. The optimum enzyme activity was exhibited at 37 °C (Figure 4b). Moreover, the addition of metal ions including $Ca^{2+}$, $Co^{2+}$, $Fe^{2+}$, $Fe^{3+}$, $K^+$, $Ni^{2+}$, $Mg^{2+}$, $Mn^{2+}$, $Cu^{2+}$, and $Zn^{2+}$ (1 mM) had no significant effect on LYR activity (Figure 4c). It has been reported that LYR is a PLP-dependent enzyme [17], and so the effect of PLP concentration was also examined. Our results showed that the addition of PLP could not further improve the specific activity of the whole-cell BL21-LYR (Figure 4d).

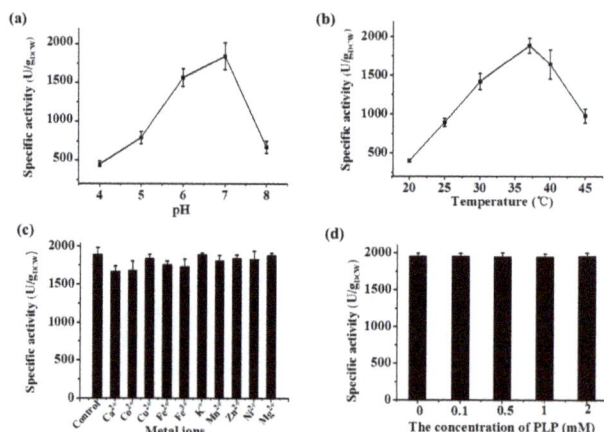

**Figure 4.** Characterization of the recombinant whole-cell BL21-LYR. (**a**) Optimal pH of the recombinant BL21-LYR; (**b**) optimal temperature of the recombinant BL21-LYR; (**c**) metal ion preference of the recombinant BL21-LYR; (**d**) the effect of PLP on the activity of the whole-cell BL21-LYR. Data are mean ± SD for three replicates.

## 2.3.2. Characterization of the Recombinant Lysine Decarboxylase

The activity of the whole-cell ATS3 containing lysine decarboxylase has been characterized in a previous study [15]. Here, the effects of pH, temperature, metal ion additive, and PLP content on the crude enzyme of ATS3 were investigated. As shown in Figure 5a, the highest specific activity of lysine decarboxylase occurred at pH 6.0, similar to the lysine decarboxylase from *E. coli* MG1655 [18]. We found that lysine decarboxylase was sensitive to pH variance. From pH 5.0 to 7.0, more than 83% of the maximum activity could be retained (Figure 5a). However, enzyme stability decreased rapidly when pH was higher than 7.0. The specific activity sharply decreased to 8.0 U/mg at pH 7.5 (Figure 5a).

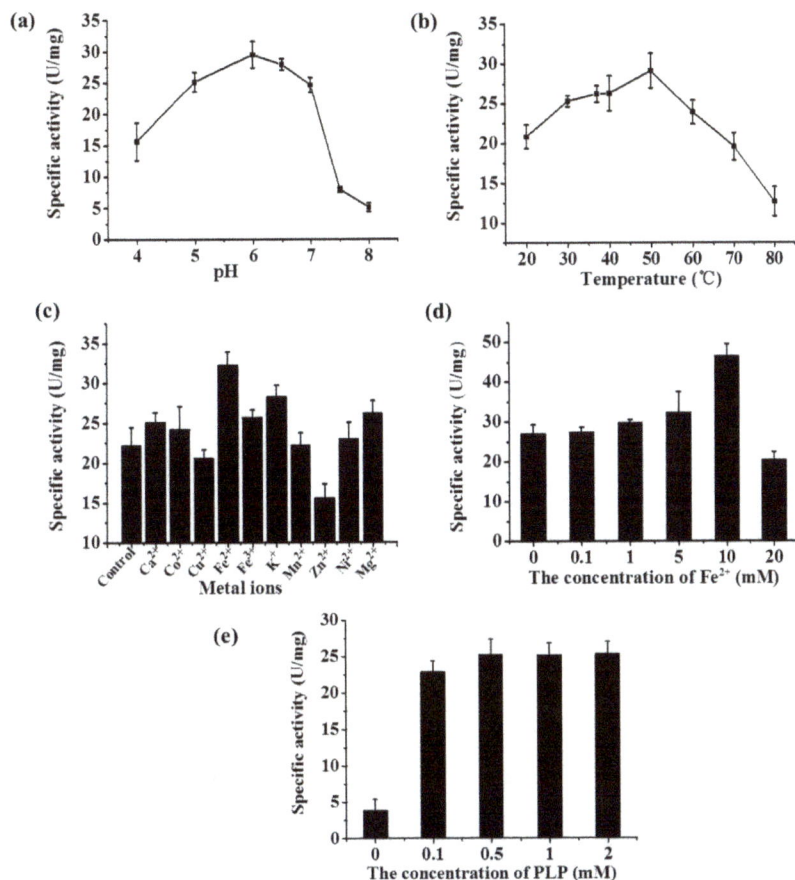

**Figure 5.** Characterization of the recombinant lysine decarboxylase. (**a**) Optimal pH of the recombinant lysine decarboxylase; (**b**) optimal temperature of the recombinant lysine decarboxylase; (**c**) metal ion preference of the recombinant decarboxylase; (**d**) optimization of the concentration of $Fe^{2+}$; (**e**) optimization of the concentration of pyridoxal 5'-phosphate (PLP). Data are mean ± SD for three replicates.

The effect of reaction temperature from 20 to 80 °C on the specific activity of lysine decarboxylase is shown in Figure 5b. The optimum activity was observed when the reaction was carried out at 50 °C. Metal ions ($Ca^{2+}$, $Co^{2+}$, $Fe^{2+}$, $Fe^{3+}$, $K^+$, $Ni^{2+}$, $Mg^{2+}$, $Mn^{2+}$, $Cu^{2+}$, and $Zn^{2+}$) (1 mM) were added into the reaction media. It was found that $Fe^{2+}$ was the optimum metal ion additive. The metal ions $Ca^{2+}$,

$Co^{2+}$, $Fe^{2+}$, $Fe^{3+}$, $K^+$, $Ni^{2+}$, and $Mg^{2+}$ also showed positive effects on lysine decarboxylase (Figure 5c). Subsequently, $Fe^{2+}$ concentration in the reaction medium was further optimized. The highest enzyme activity was obtained at a concentration of 10 mM $Fe^{2+}$ (Figure 5d), which was approximately 1.7-fold higher than that without the addition of $Fe^{2+}$.

Lysine decarboxylase is one of the PLP-dependent enzymes [5–7]. Therefore, the effect of PLP concentration in the reaction medium was examined. The specific activity of lysine decarboxylase was only 3.8 U/mg in the absence of PLP (Figure 5e). However, when 0.1 mM of PLP was added, the specific activity of lysine decarboxylase increased to 23.4 U/mg. With further increase of PLP concentration, only moderate increase was observed in enzyme activity. Taking into account the economic benefits, we determined 0.1 mM PLP to be the optimum concentration.

### 2.4. Determining the Optimal Condition of the Two-Step D-Lysine Production Process

To determine the optimal conditions of the two-enzyme cascade system, the impact factors on lysine racemase and lysine decarboxylase were compared. As has been described, the optimal pH for the whole-cell BL21-LYR was 7.0, while the specific activity of lysine decarboxylase was the highest at pH 6.0. Considering that lysine decarboxylase was more sensitive to pH variance, the two-enzyme cascade reaction was performed in potassium phosphate buffer with pH 6.0. At this pH, the racemase activity also remained at a high level. The two reactions were all carried out at around 37 °C, where the activities of the two enzymes both reached the optimum (Figures 4b and 5b). For the metal ions, in the first step, none were added into the reaction solution. After the end of the racemization reaction, 10 mM of $Fe^{2+}$ was added to improve the catalytic rate during the lysine decarboxylation process. A supplement of 0.1 mM PLP was also added during the decarboxylation reaction step.

### 2.5. D-Lysine Production at Different Substrate Concentrations

In this study, L-lysine concentrations ranging from 680 to 1710 mmol/L in the bioconversion mixture were investigated. As illustrated in Figure 6, D-lysine productivity improved with the increasing L-lysine concentration. When the initial L-lysine concentration was 680 mmol/L (Figure 6a), the production rate of D-lysine was 9.6 mmol/L/min, whereas the D-lysine production rate increased to 22.6 mmol /L/min at an L-lysine concentration of 1710 mmol/L (Figure 6d). After bioconversion for 0.5 h, all D-lysine production from different concentrations of L-lysine had nearly reached the equilibrium. After about 1 h, cell extracts of *E. coli* ATS3 were added into the reaction mixture. It was observed that the L-lysine decarboxylation process could be completed within 0.5 h. The L-lysine degradation rate was increased with increasing concentration of the resting L-lysine. Our results indicated that an initial concentration of L-lysine ranging from 680 to 1710 mmol/L appeared to have no influence on either the racemization or decarboxylation processes. As shown in Table 1, a moderate improvement in D-lysine yield was observed with an increase in L-lysine concentration. At a concentration of 680 mmol/L L-lysine, the D-lysine titer was only 287.2 mmol/L, and the molar yield was 46.9%. When the initial L-lysine concentration was increased to 1710 mmol/L, 750.7mmol/L of D-lysine was produced with a final yield of 48.8%. Meanwhile, 783.4 mmol/L cadaverine was obtained, and the total conversion yield of L-lysine was 99.7%. With the two-enzyme cascade bioconversion process we established, the enantiomeric excess of D-lysine was always higher than 99.0%.

**Table 1.** The production of D-lysine from different concentration of L-lysine.

| L-Lysine (mM) | D-Lysine (mM) | Cadaverine (mM) | D-Lysine Yield | Enantiomeric Excess |
|---|---|---|---|---|
| 680 | 287.2 | 323.5 | 46.9% | 99.5% |
| 1030 | 446.0 | 479 | 48.1% | 99.5% |
| 1370 | 601.0 | 628.5 | 48.8% | 99.4% |
| 1710 | 750.7 | 783.4 | 48.8% | 99.3% |

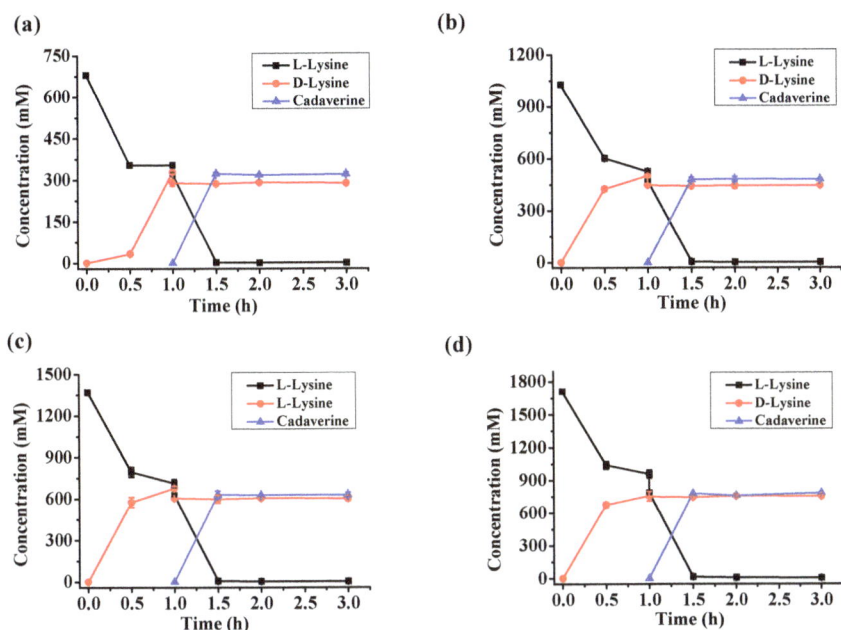

**Figure 6.** The production of D-lysine from different concentration of L-lysine. The intial L-lysine was 680 mmol/L (**a**); 1030 mmol/L (**b**); 1370 mmol/L (**c**); and 1710 mmol/L (**d**), respectively.

## 3. Discussion

D-Amino acids have been proved to be naturally present in many bacteria, plants, and animals [19–21], which exhibited widespread applications in industry, such as pharmaceutics, food, and cosmetics [5]. For example, D-amino acids can be used to synthetize the peptides which are very potent in the inhibition of HIV [22]. Thus, an efficient process for the large-scale production of D-amino acids is highly desirable. As one of the important D-amino acids, the widespread application of D-lysine in industry has also been reported [2,3]. In our study, we established a two-step process for D-lysine production from L-lysine by the microbial racemization and asymmetric degradation with lysine racemase and decarboxylase.

In our work, L-lysine was used as the starting material and racemized to generate DL-lysine under the biocatalysis of amino acid racemases. Currently, a variety of amino acid racemases, such as those for alanine, glutamate, serine, aspartate, and arginine, have been discovered in bacteria, archaea, and eukaryotes [23–25]. However, only a handful of amino acid racemases that are active with lysine have been isolated. In our study, the racemases from strain *P. mirabilis* (LYR) and *L. paracasei* (AAR) were expressed in *E. coli*. The LYR protein was highly soluble in *E. coli*, and the whole-cell BL21-LYR showed higher lysine racemization activity with an approximate D-lysine yield of 48%. The biochemical characterization study revealed that the racemase activity of the whole-cell BL21-LYR was most active at 37 °C and pH 7.0, while the characterization of purified enzyme LYR in the previous study revealed that the optimal pH was between 8.0 and 9.0, and the optimum temperature was 50 °C [8]. The high temperature and pH might affect the survival of the whole cells. Therefore, the activity of whole cell showed a difference with the purified enzyme. Meanwhile, it was found that the addition of metal ions did not enhance the activity of BL21-LYR, indicating that the racemization activity of LYR was independent of metal ions, which was in accordance with the conclusion from the previous studies on the purified enzyme LYR [8].

After the racemization reaction, the microbial asymmetric degradation of L-lysine in the DL-lysine mixture solution was performed to generate the enantiopure D-lysine. L-Lysine decarboxylase is one of the PLP fold type I enzymes, which selectively catalyze the decarboxylation of L-lysine to generate cadaverine [15,18]. The engineered *E. coli* ATS3 used in our previous study was constructed by overexpressing endogenous *cadA* gene encoding lysine decarboxylase [15,18]. Meanwhile, the ribose 5-phosphate-dependent pathway genes *pdxS* and *pdxT* from *Bacillus subtilis* were introduced into the engineered *E. coli* for de novo PLP biosynthesis [15]. As shown in Figure 3, the crude enzyme of ATS3 exhibited the highest L-lysine decarboxylation activities, and was identified as the best biocatalyst. We also determined that the lysine decarboxylase we used in this study was specific for L-lysine (Figure S1). Furthermore, the characterization of lysine decarboxylase revealed that decarboxylation activity was dependent on $Fe^{2+}$ ion and PLP additive. Bagni et al. also reported that lysine decarboxylase needs $Fe^{2+}$ as a cofactor [26]. However, the activation mechanism of $Fe^{2+}$ on decarboxylase was still unclear. For the PLP-dependent decarboxylation, it referred to the transfer of the proton during the elimination of the $CO_2$ [27]. $Fe^{2+}$ has been reported to be a good metal as redox center, which might be associated with its role in the activation of lysine decarboxylase [28].

According to the characteristics of lysine racemase and lysine decarboxylase, a two-step process for efficient D-lysine production from L-lysine was established through a two-enzyme cascade system. When L-lysine at different concentrations was used as the starting material, D-lysine with enantiomeric excess $\geq 99\%$ could always be synthesized efficiently. For example, after 1 h of racemization reaction and 0.5 h of decarboxylation reaction, 750.7 mmol/L D-lysine could be obtained from 1710 mmol/L L-lysine with a yield of 48.8%. Meanwhile, with the increase of L-lysine, the production rate of D-lysine was increased, which suggested that no substrate inhibition on the activity of whole-cell BL21-LYR was observed when the initial L-lysine concentration ranges from 680 to 1710 mmol/L. The synthesis of cadaverine was also desirable for D-lysine separation due to their large difference in the polarity. In conclusion, the two-enzyme cascade for D-lysine production described in our work shows promise for the large-scale synthesis of D-amino acids.

## 4. Materials and Methods

### 4.1. Chemicals and Enzymes

Isopropyl-β-D-1-thiogalactopyranoside (IPTG), L-lysine (>98%), and D-lysine (>98%), were purchased from Sigma Aldrich (St. Louis, MO, USA). The restriction enzymes, T4 DNA ligase and agarose gel DNA purification kit were supplied by TaKaRa Biotechnology (Otsu, Japan).

### 4.2. Bacterial Strains and Growth Conditions

The *Escherichia coli* Tans1-T1 was purchased from TransGen (Beijing, China) and used for gene cloning. The *E. coli* BL21 (DE3; TransGen) and pET-28a vector (Novagen, San Diego, CA, USA) were employed as a host for gene expression. The engineered strain *E. coli* AST3 with lysine decarboxylase (EC 4.1.1.18) activity was preserved in our laboratory [15]. The *E. coli* strains were routinely cultured in modified Luria–Bertani (LB) medium (10 g/L tryptone, 5 g/L yeast extract and 10 g/L sodium chloride) containing 50 mg/mL kanamycin (Kan).

### 4.3. Construction of Plasmids

The lysine racemase (LYR, EC 5.1.1.5) from strain *P. mirabilis* BCRC10725 and the amino acid racemase (AAR, EC 5.1.1.10) from strain *L. paracasei* ATCC 334 were reported to be capable of racemizing lysine [8,16]. The two genes were codon-optimized for *E. coli* expression system [29], and synthesized by Genewiz Company (Jiangsu, China). The two gene fragments were cloned into NcoI/XhoI sites of plasmid pET-28a to generate the plasmid pET-28a-LYR and pET-28a-AAR, respectively (Figure 2a). The two recombinant plasmids were then transformed into *E. coli* BL21

(DE3) competent cell to obtain the recombinant strain BL21/pET-28a-LYR, and BL21/pET-28a-AAR. *E. coli* BL21 (DE3) harboring empty plasmid pET-28a was used as the control.

### 4.4. Expression of Recombinant Proteins in E. coli BL21(DE3)

The recombinant strain BL21/pET-28a-LYR, and BL21/pET-28a-AAR were inoculated from a freshly transformed single colony to 10 mL of LB medium. After cultivation for 8 h at 37 °C, the recombinant strain BL21/pET-28a-LYR and BL21/pET-28a-AAR were then seeded into 100 mL of LB medium with 50 mg/L kanamycin at an inoculation volume of 1%. Cell densities were monitored by measuring the optical density at 600 nm ($OD_{600}$). Upon reaching $OD_{600}$ of 0.6~0.8, cells were induced by IPTG addition at 1.0 mM. After incubation for 12 h at 30 °C, cells were harvested by centrifugation ($6000\times g$, 5 min, 4 °C), and resuspended in 200 mM potassium phosphate buffer (pH 6.0). The recombinant strain AST3 was then seeded into 100 mL of culture medium [15] with 100 mg/L ampicillin and 34 mg/L chloramphenicol at an inoculation volume of 2%. After incubation for 12 h at 37 °C, cells were harvested by centrifugation ($6000\times g$, 5 min, 4 °C), and resuspended in 200 mM potassium phosphate buffer (pH 6.0). To prepare the crude enzyme, the cell paste was sonicated for 15 min (2 s worktime with 2 s interval). Cell extracts were then centrifuged for 15 min at $10,000\times g$ to remove cell debris. The supernatant was analyzed using sodium dodecyl sulfate-polyacrylamide gel electrophoresis (SDS-PAGE) analysis (12% acrylamide).

### 4.5. Racemase and Decarboxylase Activity Characterization

In the present study, one unit of racemase activity was defined as the amount of dry cell weight (DCW) that produced 1 μM of amino acid enantiomer from its corresponding enantiomer per minute. The impact factors associated with racemase activity were investigated in a reaction solution containing resting cells ($OD_{600}$ = 5), 820 mmol/L L-lysine, and 200 mM sodium phosphate buffer (pH = 7.0) in a total volume of 10 mL. The optimal pH for racemase activity was confirmed in the 200 mM sodium phosphate buffer pH 4.0–8.0. The effect of temperature on racemase activity was determined by measuring the enzyme activity between 20 and 45 °C. To determine the metal ions preference of racemase, the effects of $K^+$, $Fe^{2+}$, $Fe^{3+}$, $Mg^{2+}$, $Mn^{2+}$, $Cu^{2+}$, $Zn^{2+}$, $Ca^{2+}$, $Co^{2+}$, and $Ni^{2+}$ additives were investigated.

The lysine decarboxylase assay was carried out in a 50 mL centrifuge tube with 5 mL reaction broth containing 270 mmol/L L-lysine, 200 mmol/L sodium phosphate buffer (pH = 7.0), 1.0 mmol/L PLP, and 0.2 g/L crude enzyme. One unit of lysine decarboxylase activity was defined as the amount of enzyme that transformed 1 μM of L-lysine to cadaverine per minute. For pH optimization, the reaction was performed in 200 mmol/L sodium phosphate buffer (pH 4.0–8.0). To optimize temperature, the decarboxylase activity was measured in the condition with temperatures varying between 20 and 45 °C. To determine the effect of metal ions on enzyme activity, we measured decarboxylase activity in the reaction broth containing $K^+$, $Fe^{2+}$, $Fe^{3+}$, $Mg^{2+}$, $Mn^{2+}$, $Cu^{2+}$, $Zn^{2+}$, $Ca^{2+}$, $Co^{2+}$, and $Ni^{2+}$. The PLP content in the reaction mixture was also optimized.

### 4.6. The Microbial Production of Enantiopure D-Lysine and Cadaverine

In this study, L-lysine was first biologically racemized to give DL-lysine under the biocatalysis of lysine racemase. The recombinant strain BL21/pET-28a-LYR was resuspended into 20 mL of reaction mixture containing 550 mmol/L, 680 mmol/L, 1030 mmol/L, 1370 mmol/L, or 1710 mmol/L L-lysine and 200 mmol/L sodium phosphate buffer (pH 6.0) with an $OD_{600}$ of 5. The whole-cell biocatalyst was carried out in a 50 mL flask at 37 °C. Samples were taken at the specific intervals to measure the concentration of L-lysine and D-lysine.

When the reaction reached the equilibrium, the BL21/pET-28a-LYR cells were removed by centrifugation. The supernatant was maintained at 80 °C for 10 min. Then the whole cell (5.0 of $OD_{600}$) or crude enzyme (0.2 g/L) of the strain ATS3 expressing L-lysine decarboxylase was added into the supernatant to selectively catabolize L-lysine to generate cadaverine and $CO_2$. The reaction mixture also contained 200 mmol/L sodium phosphate buffer (pH 6.0), 10 mmol/L $Fe^{2+}$, and 0.1 mmol/L PLP. To identify the best biocatalyst for the enantiopure D-lysine production, the activity of resting whole cells, permeabilized cells, and crude enzyme was comparatively investigated under the condition of 550 mmol/L L-lysine.

*4.7. Analysis Methods*

The D-lysine and L-lysine in the reaction mixture were analyzed via a high-performance liquid chromatography (HPLC) system (Agilent 1290 series; Agilent, Palo Alto, CA, USA) equipped with a UV–vis detector (wavelength = 254 nm) and a chirex chiral column (Chirex 3126 (D)-penicillamine 150 mm × 4.6 mm with a pre-column 30 mm × 4.6 mm, 5 μm, Waters, Milford, MA, USA). The column was maintained at 25 °C. One millimolar $CuSO_4$ 5 $H_2O$ dissolved into water/isopropanol (95:5, *v/v*) with the flow rate of 0.8 mL/min was used as the mobile phase.

The cadaverine concentrations were determined by reverse-phase high performance liquid chromatography (HPLC) using an Agilent 1290 Infinity System equipped with a fluorescence detector (FLD, G1321B, Agilent, Palo Alto, CA, USA). The cadaverine concentration was determined by precolumn dansyl chloride derivatization following a previously described procedure [16].

## 5. Conclusions

In the current work, we have developed a two-step microbial process for high-level conversion of L-lysine to D-lysine through a lysine racemase and decarboxylase cascade system. The recombinant strain BL21-LYR, used as the biocatalyst of a racemization reaction to generate DL-lysine, was constructed by expressing lysine racemase from the strain *P. mirabilis* (LYR). Subsequently, lysine decarboxylase was employed for the asymmetric degradation of L-lysine to generate cadaverine. To produce D-lysine with high enantiomeric excess, the decarboxylation activities of resting whole cells, permeabilized cells, and crude enzyme were compared, and the crude enzyme was identified as the best biocatalyst. After characterization of the lysine racemase and decarboxylase, the two-step reaction was performed under optimal condition, and 750.7 mmol/L D-lysine could be obtained from 1710 mmol/L L-lysine. The enantiomeric excess of D-lysine was higher than 99%.

**Supplementary Materials:** The following are available online at www.mdpi.com/2073-4344/6/11/168/s1, Figure S1: Determining the substrate specificity of the lysine decarboxylase.

**Acknowledgments:** This work was supported by the National Nature Science Foundation of China (Grant No. 21576134, 21390200), "863" program of China (Grant No. 2014AA021703) and the Synergetic Innovation Center for Advanced Materials.

**Author Contributions:** Kequan Chen, Pingkai Ouyang and Xin Wang conceived and designed the experiments; Xin Wang and Li Yang performed the experiments and analyzed the data; Weijia Cao and Hanxiao Ying contributed to experimental design and also critically revised the manuscript; Xin Wang wrote the paper. All authors read and approved the final manuscript.

**Conflicts of Interest:** The authors declare no conflict of interest.

## References

1. Leuchtenberger, W.; Huthmacher, K.; Drauz, K. Biotechnological production of amino acids and derivatives: Current status and prospects. *Appl. Microbiol. Biotechnol.* **2005**, *69*, 1–8. [CrossRef] [PubMed]
2. Kaminski, H.M.; Feix, J.B. Effects of D-lysine substitutions on the activity and selectivity of antimicrobial peptide CM15. *Polymers* **2011**, *3*, 2088–2106. [CrossRef]
3. Takahashi, E.; Furui, M.; Seko, H.; Shibatani, T. D-Lysine production from L-lysine by successive chemical racemization and microbial asymmetric degradation. *Appl. Microbiol. Biotechnol.* **1997**, *47*, 347–351. [CrossRef] [PubMed]
4. Liu, Y.; Jiao, Q.; Yin, X. Preparation of D-lysine by chemical reaction and microbial asymmetric transformation. *Front. Chem. Eng. China* **2008**, *2*, 40–43. [CrossRef]

5.  Gao, X.; Ma, Q.; Zhu, H. Distribution, industrial applications, and enzymatic synthesis of D-amino acids. *Appl. Microbiol. Biotechnol.* **2015**, *99*, 3341–3349. [CrossRef] [PubMed]
6.  Anastassiadis, S. L-Lysine fermentation. *Recent Pat. Biotechnol.* **2007**, *1*, 11–24. [CrossRef] [PubMed]
7.  Hernández, S.B.; Cava, F. Environmental roles of microbial amino acid racemases. *Environ. Microbiol.* **2016**, *18*, 1673–1685. [CrossRef] [PubMed]
8.  Kuan, Y.C.; Kao, C.H.; Chen, C.H.; Chen, C.C.; Hu, H.Y.; Hsu, W.H. Biochemical characterization of a novel lysine racemase from *Proteus mirabilis* BCRC10725. *Process Biochem.* **2011**, *46*, 1914–1920. [CrossRef]
9.  Kato, S.; Hemmi, H.; Yoshimura, T. Lysine racemase from a lactic acid bacterium, *Oenococcus oeni*: Structural basis of substrate specificity. *J. Biochem.* **2012**, *152*, 505–508. [CrossRef] [PubMed]
10. Sumida, Y.; Iwai, S.; Nishiya, Y.; Kumagai, S.; Yamada, T.; Azuma, M. Identification and characterization of D-succinylase, and a proposed enzymatic method for D-amino acid synthesis. *Adv. Synth. Catal.* **2016**, *358*, 2041–2046. [CrossRef]
11. Isobe, K.; Tamauchi, H.; Fuhshuku, K.-I.; Nagasawa, S.; Asano, Y. A simple enzymatic method for production of a wide variety of D-amino acids using L-amino acid oxidase from *Rhodococcus sp.* AIU Z-35-1. *Enzym. Res.* **2010**, *2010*, 567210. [CrossRef] [PubMed]
12. Yano, S.; Haruta, H.; Ikeda, T.; Kikuchi, T.; Murakami, M.; Moriguchi, M.; Wakayama, M. Engineering the substrate specificity of *alcaligenes* D-aminoacylase useful for the production of D-amino acids by optical resolution. *J. Chromatogr. B* **2011**, *879*, 3247–3252. [CrossRef] [PubMed]
13. Pukin, A.V.; Boeriu, C.G.; Scott, E.L.; Sanders, J.P.M.; Franssen, M.C.R. An efficient enzymatic synthesis of 5-aminovaleric acid. *J. Mol. Catal. B* **2010**, *65*, 58–62. [CrossRef]
14. Sakai, A.; Xiang, D.F.; Xu, C.; Song, L.; Yew, W.S.; Raushel, F.M.; Gerlt, J.A. Evolution of enzymatic activities in the enolase superfamily: *N*-succinylamino acid racemase and a new pathway for the irreversible conversion of D- to L-amino acids. *Biochemistry* **2006**, *45*, 4455–4462. [CrossRef] [PubMed]
15. Wu, H.M.; Kuan, Y.C.; Chu, C.H.; Hsu, W.H.; Wang, W.C. Crystal structures of lysine-preferred racemases, the non-antibiotic selectable markers for transgenic plants. *PLoS ONE* **2012**, *7*, 1310–1315. [CrossRef] [PubMed]
16. Ma, W.; Cao, W.; Zhang, B.; Chen, K.; Liu, Q.; Li, Y.; Ouyang, P. Engineering a pyridoxal 5′-phosphate supply for cadaverine production by using *Escherichia coli* whole-cell biocatalysis. *Sci. Rep.* **2015**, *5*, 15630. [CrossRef] [PubMed]
17. Steffen-munsberg, F.; Vickers, C.; Kohls, H.; Land, H.; Mallin, H.; Nobili, A.; Skalden, L.; van den Bergh, T.; Joosten, H.; Berglund, P.; et al. Bioinformatic analysis of a PLP-dependent enzyme superfamily suitable for biocatalytic applications. *Biotechnol. Adv.* **2015**, *33*, 566–604. [CrossRef] [PubMed]
18. Kim, H.J.; Kim, Y.H.; Shin, J.H.; Bhatia, S.K.; Sathiyanarayanan, G.; Seo, H.M.; Choi, K.Y.; Yang, Y.H.; Park, K. Optimization of direct lysine decarboxylase biotransformation for cadaverine production with whole-cell biocatalysts at high lysine concentration. *J. Microbiol. Biotechnol.* **2015**, *25*, 1108–1113. [CrossRef] [PubMed]
19. Vranova, V.; Zahradnickova, H.; Janous, D.; Skene, K.R.; Matharu, A.S.; Rejsek, K.; Formanek, P. The significance of D-amino acids in soil, fate and utilization by microbes and plants: Review and identification of knowledge gaps. *Plant Soil* **2012**, *354*, 21–39. [CrossRef]
20. Radkov, A.D.; Moe, L.A. Bacterial synthesis of D-amino acids. *Appl. Microbiol. Biotechnol.* **2014**, *98*, 5363–5374. [CrossRef] [PubMed]
21. Fujii, N. D-Amino acids in living higher organisms. *Orig. Life Evol. Biosph.* **2002**, *32*, 103–127. [CrossRef] [PubMed]
22. Welch, B.D.; VanDemark, A.P.; Heroux, A.; Hill, C.P.; Kay, M.S. Potent D-peptide inhibitors of HIV-1 entry. *Proc. Natl. Acad. Sci. USA* **2007**, *104*, 16828–16833. [CrossRef] [PubMed]
23. Kim, P.M.; Duan, X.; Huang, A.S.; Liu, C.Y.; Ming, G.; Song, H.; Snyder, S.H. Aspartate racemase, generating neuronal D-aspartate, regulates adult neurogenesis. *Proc. Natl. Acad. Sci. USA* **2010**, *107*, 3175–3179. [CrossRef] [PubMed]
24. Couñago, R.M.; Davlieva, M.; Strych, U.; Hill, R.E.; Krause, K.L. Biochemical and structural characterization of alanine racemase from *Bacillus anthracis* (Ames). *BMC Struct. Biol.* **2009**, *9*, 53. [CrossRef] [PubMed]
25. Yoshimura, T.; Esak, N. Amino acid racemases: Functions and mechanisms. *J. Biosci. Bioeng.* **2003**, *96*, 103–109. [CrossRef]
26. Bagni, N.; Creus, J.; Pistocchi, R. Distribution of cadaverine and lysine decarboxylase activity in *Nicotiana glauca* plants. *J. Plant Physiol.* **1986**, *125*, 9–15. [CrossRef]

27. Fogle, E.J.; Toney, M.D. Analysis of catalytic determinants of diaminopimelate and ornithine decarboxylases using alternate substrates. *Biochim. Biophys. Acta—Proteins Proteomics.* **2011**, *1814*, 1113–1119. [CrossRef] [PubMed]

28. Andreini, C.; Bertini, I.; Cavallaro, G.; Holliday, G.L.; Thornton, J.M. Metal ions in biological catalysis: From enzyme databases to general principles. *J. Biol. Inorg. Chem.* **2008**, *13*, 1205–1218. [CrossRef] [PubMed]

29. Condon Adaptation Tool Division Home Page. Available online: http://www.jcat.de/ (accessed on 20 November 2015).

MDPI

St. Alban-Anlage 66

4052 Basel

Switzerland

Tel. +41 61 683 77 34

Fax +41 61 302 89 18

www.mdpi.com

*Catalysts* Editorial Office

E-mail: catalysts@mdpi.com

www.mdpi.com/journal/catalysts

www.ingramcontent.com/pod-product-compliance
Lightning Source LLC
Chambersburg PA
CBHW051906210326
41597CB00033B/6041